Geodynamics of the Eastern Pacific Region, Caribbean and Scotia Arcs

Geodynamics Series

The Final Reports of the International Geodynamics Program sponsored by the Inter-Union Commission on Geodynamics.

Geodynamics of the Eastern Pacific Region, Caribbean and Scotia Arcs

Edited by Ramón Cabré, S.J.

Geodynamics Series
Volume 9

American Geophysical Union
Washington, D.C.

Geological Society of America
Boulder, Colorado

1983

Final Report of Working Group 2, Geodynamics of the Eastern Pacific Region,
Caribbean and Scotia Arcs, coordinated by C. L. Drake on behalf of the Bureau
of Inter-Union Commission on Geodynamics

American Geophysical Union, 2000 Florida Avenue, N.W.
  Washington, D.C. 20009

Geological Society of America, 3300 Penrose Place, P.O. Box 9140
  Boulder, Colorado 80301

Library of Congress Cataloging in Publication Data

Main entry under title:

Geodynamics of the eastern Pacific region, Caribbean and Scotia arcs.

   (Geodynamics series, ISSN 0277-6669; v. 9)
   Bibliography: p.
   1. Plate tectonics--Addresses, essays, lectures. 2. Geology--Latin
   America--Addresses, essays, lectures.
I. Cabré, Ramón. II. Series.
QE511.4.G455   1983   551.1'36   83-3703
ISBN 0-87590-502-1

Printed in the United States of America

CONTENTS

FOREWORD

After a decade of intense and productive scientific cooperation between geologists, geophysicists and geochemists the International Geodynamics Program formally ended on July 31, 1980. The scientific accomplishments in more than seventy scientific reports and in this series of Final Report volumes.

The concept of the Geodynamics Program, as a natural successor to the Upper Mantle Project, developed during 1970 and 1971. The International Union of Geological Sciences (IUGS) and the International Union of Geodesy and Geophysics (IUGG) then sought support for the new program from the International Council of Scientific Unions (ICSU). As a result the Inter-Union Commission on Geodynamics was established by ICSU to manage the International Geodynamics Program.

The governing body of the Inter-Union Commission on Geodynamics was a Bureau of seven members, three appointed by IUGG, three by IUGS and one jointly by the two Unions. The President was appointed by ICSU and a Secretary-General by the Bureau from among its members. The scientific work of the Program was coordinated by the Commission, composed of the Chairmen of the Working Groups and the representatives of the national committees for the International Geodynamics Program. Both the Bureau and the Commission met annually, often in association with the Assembly of one of the Unions, or one of the constituent Associations of the Unions.

Initially the Secretariat of the Commission was in Paris with support from France through BRGM, and later in Vancouver with support from Canada through DEMR and NRC.

The scientific work of the program was coordinated by ten Working Groups.

WG 1 Geodynamics of the Western Pacific-Indonesian Region
WG 2 Geodynamics of the Eastern Pacific Region, Caribbean and Scotia Arcs
WG 3 Geodynamics of the Alpine-Himalayan Region, West
WG 4 Geodynamics of Continental and Oceanic Rifts
WG 5 Properties and Processes of the Earth's Interior

WG 6 Geodynamics of the Alpine-Himalayan Region, East
WG 7 Geodynamics of Plate Interiors
WG 8 Geodynamics of Seismically Inactive Margins
WG 9 History and Interaction of Tectonic, Metamorphic and Magmatic Processes
WG 10 Global Syntheses and Plaeoreconstruction

These Working Groups held discussion meetings and sponsored symposia. The papers given at the symposia were published in a series of Scientific Reports. The scientific studies were all organized and financed at the national level by the national committees even when milti-national programs were involved. It is to the national committees, and to those who participated in the studies organized by those committees, that the success of the Program must be attributed.

Financial support for the symposia and the meetings of the Commission was provided by subventions from IUGG, IUGS, UNESCO and ICSU.

Information on the activities of the Commission and its Working Groups is available in a series of 17 publications: Geodynamics Reports, 1-8, edited by F. Delany, published by BRGM; Geodynamics Highlights, 1-4, edited by F. Delany, published by BRGM; and Geodynamics International, 13-17, edited by R. D. Russell. Geodynamics International was published by World Data Center A for Solid Earth Geophysics, Boulder, Colorado 80308, USA. Copies of these publications, which contain lists of the Scientific Reports, may be obtained from WDC A. In some cases only microfiche copies are now available.

This volume is one of a series of Final Reports summarizing the work of the Commission. The Final Report volumes, organized by the Working Groups, represent in part a statement of what has been accomplished during the Program and in part an analysis of problems still to be solved. This volume from Working Group 2 was edited by Ramon Cabre, S. J.

At the end of the Geodynamics Program it is clear that the kinematics of the major plate movements during the past 200 million years is well understood, but there is much

1

less understanding of the dynamics of the processes which cause these movments.

Perhaps the best measure of the success of the Program is the enthusiasm with which the Unions and national committees have joined in the establishment of a successor program to be known as:
Dynamics and evolution of the lithosphere:

The framework for earth resources and the reduction of the hazards.

To all of those who have contributed their time so generously to the Geodynamics Program we tender our thanks.

C. L. Drake, President, ICG, 1971-1975
A. L. Hales, President, ICG, 1975-1980

INTRODUCTION

Ramon Cabré, S.J., Chairman of WG2,

Observatorio San Calixto, La Paz, Bolivia

Geodynamics in the Region of WG2

Working Group No. 2, one of the four 'regional' Groups of the Geodynamics Project, has been concerned with the Eastern Pacific and Western Americas and with the Caribbean and Scotia Arcs. This has represented an exceptionally active and diverse area of study, diverse in the factors conditioning research, such as climates (ranging from the western United States to the poorest countries of Latin America) and the geodynamic conditions revealed in the interior of the earth.

The region studied by WG2 included the western continental margin of North and South America or in plate tectonic terms, the region dominated by the interaction between the oceanic plates of the Pacific Ocean and the continental plates of the Americas. Here some of the highest velocities of convergence on the globe have resulted in the formation of the remarkable Cordilleran Arcs.

Deep seismic activity marks present subduction to a depth of at least 600 km beneath central South America and along about half the length of the continent. Activity to intermediate depths occurs in the remainder of South America (except the extreme south) and in Central America, the Caribbean and Scotia Arcs and closely traces the extent and dimensions of subduction in these regions. North America, except Alaska, is characterised by the absence of seismic activity below 70 km - in California because there is no current subduction and in the Pacific North-West, because the Juan de Fuca plate is young and hot and becomes plastic at a shallow depth.

Although the varying size of the plates studied by WG2, from the giant Pacific plate to the fragmented Juan de Fuca plate, has been another diversifying factor, similarities have begun to emerge. For example, the Caribbean and Scotia Arcs were apparently produced by the same process but are at different stages. They offer weak resistance to the eastward push of Pacific ocean plates resulting in arc displacement and basin formation.

Collaboration among Researchers

The research intensity applied to this range of problems has varied considerably according to the facilities and support available in each country. In this respect, the interest and generosity of the host of people and institutions assisting in the research programs of less well endowed regions and remote areas deserves special mention.

In particular we would like to remember John D. Weaver who died as the Geodynamics Project was approaching its end. He contributed with enthusiasm as Chairman of the Caribbean Study Group and as volunteer translator for parts of this volume written in Spanish. Unfortunately he died before he could complete this work.

We would also like to mention three Institutions whose contributions of interest and encouragement were particularly important: Department of Terrestrial Magnetism of the Carnegie Institution of Washington, Pan American Institute of Geography and History, The Scientific Program of the Organisation of American States. These are only a few of many, we could continue with an almost endless list of Universities and other Institutions. In many cases, such as the Scotia Arc, cooperation is the 'conditio sine qua non' for research; in the work of WG2, co-operation has been consistently evident.

Constitution and Activity of WG2

The following membership of WG2 was initially selected:

Allen, C.R., Seismological Lab., California Institute of Technology. Pasadena.
Barker, P.F., Dept. of Geophysics, Univ. of Birmingham. U.K.
Cabré, T., Observatorio San Calixto. La Paz, Bolivia. Chairman.
Casaverde, M., Instituto Geofisico del Perú. Lima, Perú.
Chase, R.L., Dept. of Geology, Univ. of

British Columbia. Vancouver, Canada.

Dengo, G., ICAITI. Guatemala.

Fiedler, G., Inst. Seismológico, Observ. Cagigal. Caracas, Venezuela.

Gabrielse, H., Dept. of Energy, Mines and Resources. Vancouver, Canada.

Gonzáles-Ferrán, O., Depto. Geologia, Univ. de Chile. Santiago, Chile.

Linares, E., Depto. Ciencias Geológicas, Univ. de Buenos Aires. Argentina.

Lomnitz, C., Inst. de Geofísica, UNAM. México.

Ramírez, J.E., Inst. Geofísico de los Andes Colombianos. Bogotá, Colombia.

Ritsema, A.R., Koninlijk Nederlands Meteoro- logisch Instituut. De Bilt, Netherlands.

Tomblin, J., Seismic Research Unit, Univ. of the W. Indies. St. Augustine, Trinidad.

Van Andel, Tj. R., Dept. of Oceanography, Oregon State Univ. Corvallis, Oregon.

Weaver, J., Dept. of Geology, Univ. of Puerto Rico. Mayaguez, Puerto Rico.

Several changes were later made in WG2 membership at different times and for different reasons: Allen, Chase, Fiedler, Linares, Lomnitz, Ritsema and Van Andel left the WG2; in their place the following researchers entered the Group:

Baldis, B.A., Museo Argention de Ciencias Naturales. Buenos Aires, Argentina.

Beets, D.J., Geological Institute. Amsterdam, Netherlands.

Bellizzia, A., Servicio Geológico Nacional, Ministerio de Minas y Emergía. Caracas, Venezuela. Co-Chairman.

Del Castillo, L., Instituto de Geofísica. UNAM. México.

Dorman, L.R.M., Scripps Inst. of Oceanography. La Jolla, California.

Kulm, L.V., Dept. of Oceanography, Oregon State Univ. Corvallis, Oregon.

Riddihough, R.P., Pacific Geoscience Centre. Sidney, B.C., Canada.

Shimamura, H., Dept. of Geophysics, Hokkaido Univ. Saporo, Japan.

WG2 was faced, among other difficulties, with a considerable communications problem. Distances and lack of funding on the one hand and overlapping commitments on the other, challenged mail, meeting and press communication. Meetings were held in: Lima (August 22-24, 1973); Vancouver (August 19-20, 1975); Acapulco (June 14-15, 1976); Curacao (July 22-23, 1977); Bogotá (March 28-30, 1979).

To stimulate communication between the WG2 members and with the geoscience community, the Co-Chairman of the WG2 edited a "Boletín de Geodinámica" in Spanish, the language spoken in the part of the region where communication was most difficult. It was financed by Venezuelan institutions. Two issues were published.

Study Groups

Because it became clear that a single coordinating group acting over so large a region would be practically impossible, zone study groups were organized. This had the added advantage of involving a larger number of geoscientists:

Study Group	Chairman
Juan de Fuca Plate	Riddihough (initially, Chase)
Cocos Plate	Del Castillo
Caribbean	Weaver (initially, Dengo)
Nazca Plate	Kulm (initially, Lomnitz)
Scotia Arc	Gonzáles-Ferrán
Argentina Comm. on Andean Geodynamics	Baldis

The results of the Study Groups have been uneven. This is not to say that zones assigned to less active Groups were neglected, but that in practice much research effort was often coordinated under National Geodynamic Committees (for instance in the United States). This explains, at least partially, why the present volume presents geodynamic research and results in a varied manner. The different studies are presented in geographical order (both Study Group and National Reports). [Note that the responsibility for the individual reports in this volume rests entirely with the listed authors.]

Summary Comments

Although advances in geodynamics have been important, possibly the most significant result of the Geodynamics Project in the region has been the increase in communication between geoscientists. There is still, however, considerable room for improvement. In the subject of our research, a number of major geodynamic problems remain. Some of these problems existed before the Project began, others have appeared as a consequence of new discoveries. Clearly there is ample opportunity and need for a continuing effort, an effort that could possibly be coordinated through a similar project.

Acknowledgements. Final revision of this introduction by Robin Riddihough is acknowledged and appreciated.

GEODYNAMICS OF THE JUAN DE FUCA PLATE

R.P. Riddihough[1], M.E. Beck[2], R.L. Chase[3], E.E. Davis[1],
R.D. Hyndman[1], S.H. Johnson[4] and G.C. Rogers[1]

Abstract. The Juan de Fuca plate in the N.E.
Pacific Ocean, although small by global
standards, is very complex. Over the last 10 m.
years, the Juan de Fuca ridge at which it has
been generated, has slowed its spreading rate
(from 8 cm/yr to below 6 cm/yr), has become
segmented and has rotated clockwise. This
process has been associated with fragmentation
into a series of sub-plates, each with its own
recent history. Oblique convergence with the
margin of North America at 2-4 cm/yr is a
consequence of plate geometry and is supported
by deformation of the sediments of the
continental slope and the burial of sea-floor
magnetic anomalies. Low heat flow, recent
movements and the regional gravity field of
western British Columbia and the NW United States
provide some characteristic features of a
subduction zone, as does the calc-alkaline volcanic chain of the Mt. Garibaldi - Cascades
system. Although the lack of a deep Benioff zone
is explicable as a function of the slow
subduction of a young plate, the absence of any
seismicity specifically associated with underthrusting remains unexplained. Amongst the
other features which are now well established
but not yet understood in geodynamic terms are
regional north-south compression in the adjacent
land areas, the rotations of blocks within the
arc-trench gap, and the origin and influence of
the large seamount province immediately west of
the ridge system.

Introduction

The Juan de Fuca plate has occupied a unique
place in the development of recent geodynamics
studies in that the magnetic surveys in the NE

Pacific Ocean by Raff and Mason (1961) (Figure 1)
provided a critical piece of evidence which led
to the development of sea floor spreading
theory and plate tectonics. Key papers by
Wilson (1965), Vine and Wilson (1965) and Tobin
and Sykes (1968) followed by the work of Atwater
(1970) established that the Juan de Fuca ridge
was a spreading ridge and that between the ridge
and the North American continent lay a separate
plate which was probably subducting along the
continental margin.

By the beginning of the Geodynamics Project
the main features of the area had been established (Dehlinger et al., 1971); however it was
widely recognized that the area was extremely
complex and that many of the accepted criteria
of the simple plate tectonic model were missing
or unrecognizable. Despite Atwater's exposition
of the implications of plate tectonic modelling
for the North-east Pacific, the climate of
scientific opinion was not whole-heartedly in
favour of its application to the region. In
our opinion, one of the main achievements of the
work done during the period of the Geodynamics
Project has been to gain wide acceptance for the
existence of the Juan de Fuca plate and to
develop a realization that the apparent
contradictions of its geodynamics are not so
much proof that plate tectonics does not work in
this region but examples of the detailed
complexity of plate tectonics in action.

The Ridge

The Juan de Fuca ridge system consists of a
number of different segments (Figs. 2 and 3).
From south to north, these are the Gorda ridge,
the Juan de Fuca ridge, the Explorer ridge and
the Dellwood knolls. The main segments are
connected by right-lateral transform faults
which shift the ridge north of the transform to
the west. The transforms were initiated at
progressively later times from south to north:
Mendocino fracture zone (70 m.y.), Blanco
fracture zone (15 m.y.), Sovanco fracture zone
(8 m.y.) and the Revere-Dellwood fracture zone
(1-2 m.y.). Rates of motion across these
transforms are at present of the order of
5 cm/yr but in detail the individual sections

[1]Pacific Geoscience Centre, Earth Physics Branch,
Dept. Energy, Mines and Resources, Box 6000,
Sidney, B.C., Canada, V8L 4B2
[2]Geology Dept., Western Washington Univ.,
Bellingham, Washington 98225
[3]Geology Dept., Univ. of British Columbia,
Vancouver, B.C., Canada
[4]School of Oceanography, Oregon State Univ.,
Corvallis, Oregon 97331

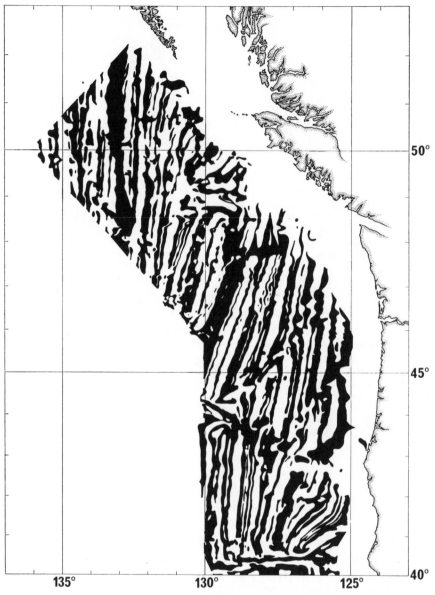

Fig. 1. Magnetic anomaly map of the north-east Pacific Ocean from Raff and Mason (1961). Shaded areas — anomalies > 0.

of the ridge seem to be spreading at differing rates.

Close analysis of magnetic anomalies and their spacing has become possible through what is virtually a complete magnetic coverage over the whole Juan de Fuca plate and ridge (Raff and Mason, 1961; Emilia et al., 1968; Srivastava et al., 1971; Potter et al., 1974; Macleod et al., 1977; Tiffin and Riddihough, 1977). Using the results from global studies on the detailed chronology of magnetic reversals, it has become clear that over at least the last 10 million years there has been a general decrease in spreading rate along the system (i.e. Explorer ridge, 7 - 4 cm/yr; Juan de Fuca ridge, 8 - 6 cm/yr; Gorda ridge north, 8 - 6 cm/yr, Gorda ridge south, 8 - 3 cm/yr; Atwater and Mudie, 1973; Klitgord et al., 1975; Riddihough, 1977). At various times back to 30 m.y., spreading has been asymmetric, with faster spreading predominantly occurring on the eastern limb of the ridge (Elvers et al., 1973, Riddihough, 1977). Variable spreading rates have also been involved in the progressive clockwise rotation of the axes of all the Juan de Fuca ridge segments. The major period of rotation was between 3 and 5 m.y.b.p. and the amount of rotation from 10-15°.

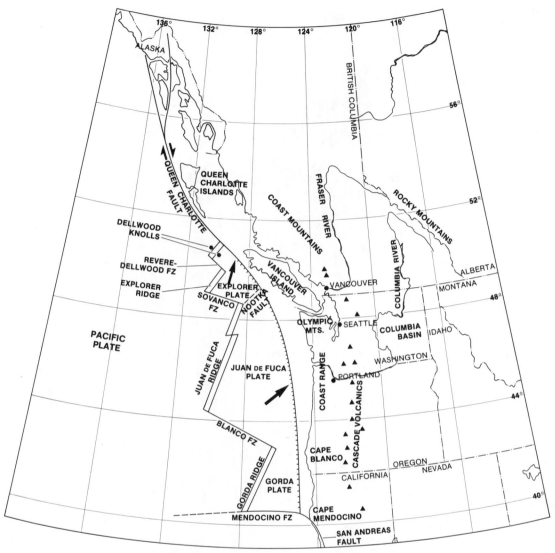

Fig. 2. Location map of the Juan de Fuca plate and surrounding areas. Arrows on Explorer and Juan de Fuca plates indicate movement relative to N. America continent, triangles are Quaternary volcanoes.

The morphology and tectonics of the ridge differ considerably along the length of the system (Fig. 4). The principle contrasts are between sections of the ridge with a pronounced axial valley and sections consisting of a broad ridge with little or no axial valley (Davis and Lister, 1977). The Gorda ridge which has a very slow spreading rate along its southern section (around 2 cm/yr), has an axial valley (Atwater and Mudie, 1973) as do sections immediately to the south of the intersection of the Sovanco and Revere-Dellwood fracture zones. Approximately half of the length of the Juan de Fuca ridge system has little or no axial valley, a situation which may reflect its intermediate status between a slow and a fast spreading ridge system.

The detailed structure of the axial valley sections of the ridge has been made more visible by considerable rates of turbidite sedimentation. Clear sedimentary sections have allowed the pattern of inward dipping normal faults with uplifted and rotated fault blocks away from the valley to be clearly seen and evaluated (Davis et al., 1976, Davis and Lister, 1977). At the northern end of the Juan de Fuca ridge, fault blocks range from 1 - 8 km in width and have been rotated up to 8°. Fault dips range from 45° to near vertical and individual horizontal displacements are estimated at near 0.1 cm/yr. This is an order of magnitude too slow to satisfy the estimated spreading rate at this section of the ridge (5 - 8 cm/yr) and supports the identi-

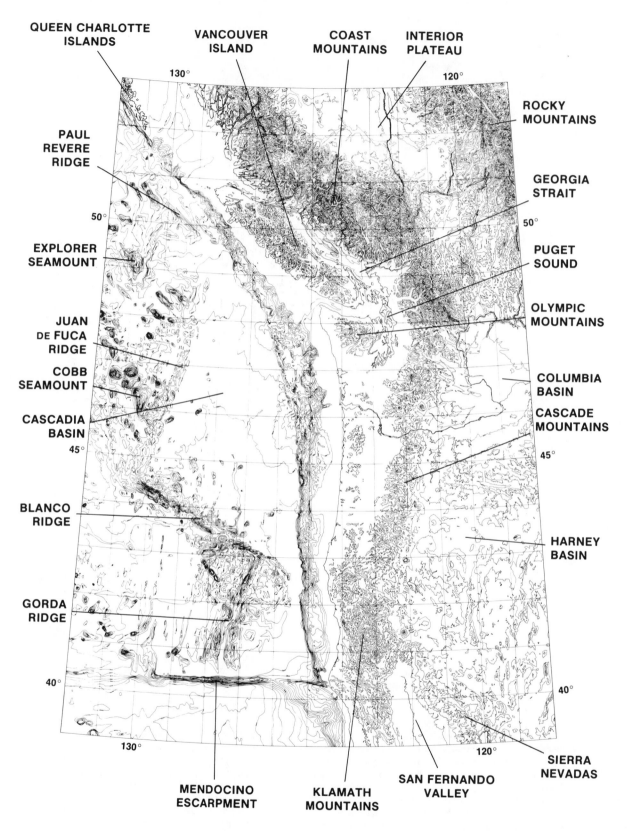

Fig. 3. Relief map of Juan de Fuca plate and its zone of interaction with the N. American continent.

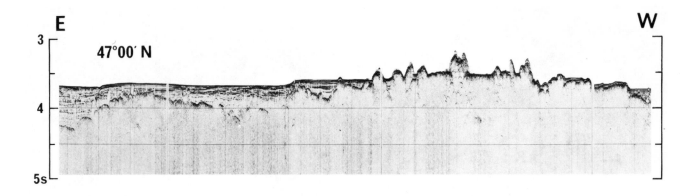

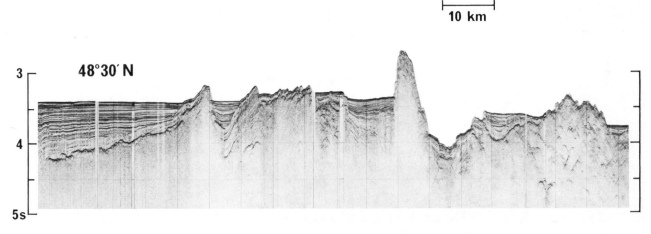

Fig. 4. Reflection seismic profiles across the Juan de Fuca ridge at 48°30'N and 47°N showing segments where the rift valley is developed and where it is absent.

fication of a twin spreading valley system in this area (Barr and Chase, 1974). A twin valley system, the western valley being the most recent, is also seen where the Explorer ridge meets the Revere-Dellwood fracture zone (Srivastava et al., 1971; Hyndman et al., 1978).

More than 300 heat flow measurements have been made along the Juan de Fuca ridge system. Although these measurements show a general decrease in heat flow with the increasing age of the sea-floor (Korgen et al., 1971), their variability has led to a closer investigation of the relation between heat-flow and sedimentation (Davis and Lister, 1977; Hyndman et al., 1978). From those studies it is clear that heat flow values are strongly influenced by the circulation of sea water in the fractured crust and that true estimates of total heat flux may only be possible where thick overlying sediments inhibit convection heat loss. For observations in such areas near the Juan de Fuca system however, values are still predominantly below the theoretical levels predicted from sea-floor age (Hyndman et al., 1978; Davis et al., 1979).

Seismicity on the Juan de Fuca ridge system (Fig. 5), is apparently confined to the major transforms and to the sections of the ridge exhibiting an axial valley. The section of the Juan de Fuca ridge between 48°N and 45°N exhibits little or no seismicity (Tobin & Sykes, 1968; Chandra, 1974). A persistent problem in the understanding of offshore seismicity has been the lack of close correlation between epicentral plots based on land data and tectonic features as determined by marine geophysical surveys (Milne et al., 1978). Three sections of the Juan de Fuca system have been investigated in more detail using deployments of either ocean bottom seismometers or sonobuoys. These investigations (Johnson and Jones, 1978; Jones and Johnson, 1978; Hyndman et al., 1978) showed that mislocations of epicentres from land stations of up to 60 km were likely and that in detail, correlation of micro-seismicity with bathymetric and tectonic features is good.

Focal mechanisms along the ridge-transform system (Chandra, 1974; Milne et al., 1978; Wetmiller, 1971, Tatham and Savino, 1974)

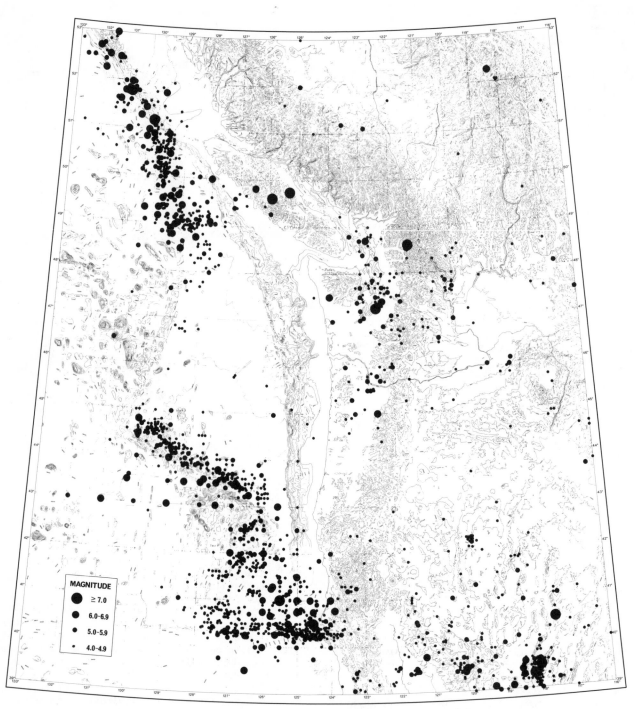

Fig. 5. Seismicity of the Juan de Fuca plate, all events, magnitude ≥ 4 to the end of 1977.

MAGNITUDE

● ≥ 7.0

● 6.0-6.9

● 5.0-5.9

· 4.0-4.9

(Figure 6) almost all indicate strike-slip faulting with a probable northwest southeast orientation. Despite the structural evidence within the axial valley, only one solution indicates normal faulting (Jones & Johnson, 1978).

The Plate System

East of the ridge, the seafloor is underlain by the Juan de Fuca plate, the descendant of a much older plate called the Farallon plate by Atwater (1970). This plate is now regarded as

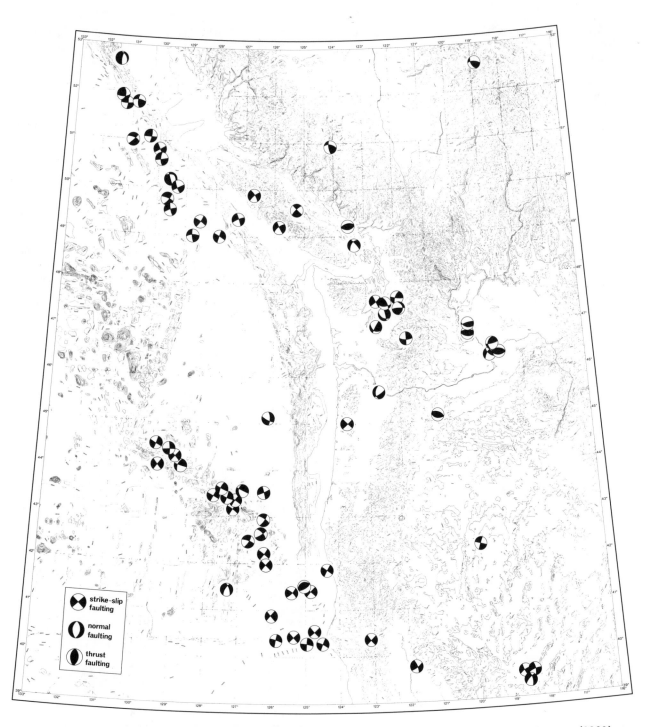

Fig. 6. Fault plane solutions in the Juan de Fuca plate region. For sources see Rogers (1980).

composed of a series of sub-plates attached to the various segments of the Juan de Fuca ridge. The motion of these plates relative to the Pacific plate is complementary to the history of the spreading ridge system and has presumably also changed over the last few million years. The assumption that plate motion has either been perpendicular to the strike of the ridge or parallel to the active transforms (a difference of about 10°) seems to fit the majority of tectonic and geometric models (e.g. Atwater and Mudie, 1973; Riddihough, 1977). From the azimuth of magnetic anomalies, plate rotation seems to have accompanied the observed ridge rotation in most cases and has been clockwise from 10 - 20° during various periods in the last 5 m.y. The southern part of the Gorda plate may have rotated by up to 40° within a short period around 3 m.y.b.p. Of note is the fact that prior to 5 m.y.b.p., at least as far back as 40 m.y.b.p., the ridge system seems to have retained a stable north-south direction. A recent change in the geodynamic framework that controls the system seems to be implied.

The independence of the sub-plates created by the differing segments of the ridge has resulted in at least one major left lateral, northeast oriented fracture zone (plate boundary) which extends from the ridge to the continental margin (Fig. 7). The boundary, between the Explorer and Juan de Fuca plates, is called the Nootka Fault (Hyndman et al., 1978). It is clearly active (Milne et al., 1978) and because of continual adjustment to changing plate motions has had a very complex history (Riddihough, 1977). The present rate of displacement across it is estimated to be about 3 cm/yr.

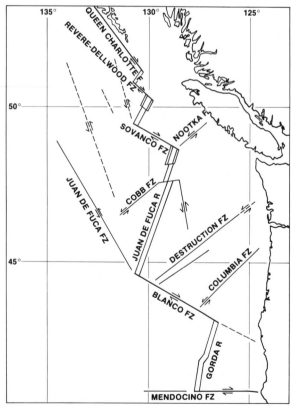

Fig. 7. Fracture zones in the Juan de Fuca plate. Arrows show movement on ridge-ridge transforms and apparent displacement on oblique fractures.

Further south, two other apparently similar fracture zones have left their trace in the sea-floor magnetic anomalies and although not active now, are estimated to have moved left-laterally at rates of up to 3.5 cm/yr (Silver, 1971). Between the Gorda plate and the main area of the Juan de Fuca plate, no active fracture has been identified. As spreading rates north and south of the Blanco fracture zone are similar, such a zone may not be geometrically necessary. Nevertheless seismicity within the Gorda plate has at various times (e.g. Seeber et al., 1970; Herd, 1978) been correlated with extensions of the San Andreas Fault.

The main change in character along the Gorda Ridge, takes place near 41°30'N, where a northerly section of the ridge currently spreading at 6 cm/yr is contrasted to a southern section spreading at 2.4 cm/yr. (see Fig. 9). Distortion of the magnetic anomalies in the southern part of the Gorda plate (Silver, 1971) indicates that internal deformation has taken place within the last 5 m.y. in this area and the present seismicity suggests that it may be continuing. Fault plane solutions are indeterminate between NW-SE and SW-NE strike-slip motions although present opinion favours NE trending, left-lateral solutions (Silver, 1971; Smith and Knapp, 1979).

Because of the damming effect of the Juan de Fuca ridge system and high rates of sedimentation from the surrounding land area (von Heune and Kulm, 1973) most of the Juan de Fuca plates are covered by 1-3 km of sediment. The original refraction work of Shor et al. (1968) gave sediment thickness of 1.3 to 2.5 km across the Cascadia Basin with total crustal thicknesses of between 6 and 8 km. The crustal thickness of the Gorda plate was thinner at around 4 km. Reflection data (McManus et al., 1972; Seely, 1977) magnetic anomaly depth determinations (Barr, 1974) and Deep Sea Drilling Results (von Heune and Kulm, 1973) have confirmed the sediment thicknesses. Further refraction surveys have shown that the Explorer plate may be thicker than the other plates of the system (Malacek and Clowes, 1978).

Hot Spots and Sea-mounts

The role of hot spots in the dynamics of the NE Pacific has so far eluded clear exposition. The sea floor west of the Juan de Fuca ridge system is characterized by a large number of sea mount chains (Figure 8) with a consistent north-westerly trend. Many of these are short, apparently generated at the spreading ridge and are of the same age as the ocean floor on which they are now located (Barr, 1974). The probable motion of the Juan de Fuca plate over the mantle suggests that similar seamounts east of the ridge would have a north-south orientation (Silver et al., 1974). However, the striking lack of any seamount chains on the Juan de Fuca plate is not yet understood.

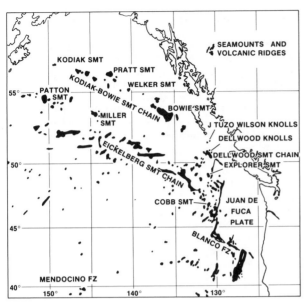

Fig. 8. Seamounts and volcanic ridges in the north-east Pacific Ocean.

Amongst the seamounts are two major chains which appear to be related to a fixed mantle plume - the Eikelberg chain and the Kodiak-Bowie chain. These stretch up to 1500 km northwestwards across the Pacific plate and although not yet closely studied, seem to be younger towards the Juan de Fuca ridge system (Silver et al., 1974; Vogt and Johnson, 1975; Vogt and Byerly, 1976). The age of progression and orientation of the Kodiak-Bowie seamount chain in particular, is consistent with other estimates of Pacific plate motion. The youngest part of the chain is thought to be the Tuzo Wilson Knolls close to the continental margin between the Queen Charlotte Islands and northern Vancouver Island and within 70 km of the most northerly spreading segment of the ridge system (Chase, 1977). It has been speculated that a fixed mantle volcanic source near or east of this point might influence both the dynamics of the ridge system and the structure on the continental margin. The eastern end of the Eickelberg chain, near Cobb seamount, has also been suggested as having an influence on the history and development of this portion of the ridge (Vogt and Byerly, 1976).

Possible hotspot traces east of the Juan de Fuca plate are the east-west Massett-Anahim belt in Canada (Bevier et al., 1979) and the Snake River - Yellowstone belt in the U.S.A. (Smith et al., 1977).

Major Triple Junctions

North end of the plate

The northern end of the Juan de Fuca ridge spreading system is a triple junction of the

ridge-fault-trench type (Fig. 9). The history of the position of this triple junction was outlined by Atwater (1970) and in terms of plate tectonic theory is dependent upon the relative motions of the three plates involved. Stability calculations (Chase et al., 1976) and reconstructions from magnetic anomaly positions (Riddihough, 1977) agree that using the best estimates for Pacific/America and Pacific/Explorer plate motions, the triple junction remained more or less stable off northern Vancouver Island from 10 to 4 m.y.b.p. It then began to migrate northwestwards at rates of the order of 1-2 cm/yr. Independent geological evidence for the stable position of the triple junction has been shown by the change in the nature of the deformation of structures beneath the continental slope from compression to normal or strike-slip faulting (Tiffin et al. 1972).

The present position of the triple junction seems to be near the Dellwood knolls, a short spreading segment joined to the Explorer ridge by the Revere-Dellwood fracture zone (Fig. 9).

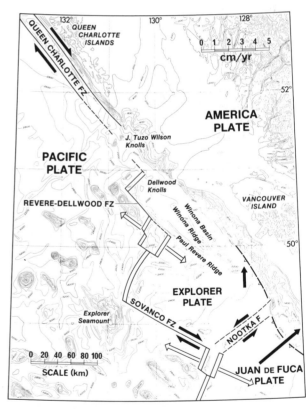

Fig. 9. Features and movements of the Explorer plate and northern triple junction region. Arrows represent motion vectors in cms/yr calculated from magnetic anomalies. Split arrows are transform/strike slip motion, half the motion on each side. Solid arrows at the margin are motion relative to continent and N. America plate. Open arrows are ridge spreading, half the motion on each side of the ridge.

Although these two knolls lack many of the topographic features which are characteristic of the ridge system to the south, dredge sampling, heat-flow, seismicity, and reflection seismic data (Srivastava et al., 1971; Hyndman et al., 1978) strongly suggest that they are the locus of active spreading. Recent sedimentation rates in the area are high (up to 10 cm per 1000 yrs) so that the precise junction with the Queen Charlotte Fault has not yet been located.

South end of plate

The triple junction at the southeastern corner of the Gorda plate is of the fault-fault-trench type bringing together the San Andreas fault system, the Mendocino fracture zone and a convergence zone along the continental margin of northern California and southern Oregon (Fig. 10). The tectonics of this junction are complicated by the fact that the Pacific plate to the south includes a block of continental material and that the junction is moving north-

northwestwards with respect to the American plate at a rate of 5-6 cm/yr.

The Mendocino fracture zone (one of the major fracture zones of the N. Pacific ocean, initiated 70 m.y.b.p.) is marked by a strong topographic ridge. Its active portion is approximately 250 km long with an azimuth of 90°. This is perpendicular only to the most recent, southern section of the Gorda ridge so that unless spreading has been oblique to the ridge, geometry demands a considerable amount of earlier convergence beneath the Mendocino Ridge (Silver, 1971). Such convergence is in accord with seismic and gravity interpretations (Dehlinger et al., 1971) but not the majority of present fault plane solutions (Couch, 1979).

Recent convergence of the Gorda plate along the continental margin is shown by sedimentary structures and magnetic data (Silver, 1971) and by the existence of an easterly dipping zone of seismicity (Smith and Knapp, 1979). However, direct seismic evidence of present underthrusting processes is absent and the plate seems to be fracturing in place in response to a general north-south compression. Detailed mapping of the geology on land in the immediate area of the triple junction has shown that there is a considerable zone of active faults north of Cape Mendocino which are sub-parallel with the San Andreas fault and which may bound slivers of material adjusting to the northward movement of the junction (e.g. Herd, 1978).

Convergence at the Continental Margin

The convergence of the Juan de Fuca plate system with the America plate is a geometrical consequence of the vector solution of the Juan de Fuca/Pacific and Pacific/America plate motions. This was first shown by Atwater (1970) to result in oblique convergence (Azimuth 40-50°) of between 3-4 cm/yr. That such convergence has taken place at least up to 6 m.y. ago is shown by the dating of the youngest buried magnetic anomaly in the ocean crust below the continental shelf off Vancouver Island (Barr, 1974).

Although the history of Juan de Fuca/Pacific plate motions is well established over the last 10 m.y. from the ridge history reviewed above, the recent history of Pacific/American motions is very dependent upon key interpretations of spreading in the Gulf of California (e.g. Larson, 1972) and movements along the San Andreas fault system. Our understanding of this motion may change with future investigations; however all present 'best estimate' calculations result in convergence between the America and Juan de Fuca plate systems at similar rates to those suggested by Atwater (1970) (e.g. Silver, 1971; Chase et al., 1976; Riddihough, 1977).

One critical piece of evidence which has been cited against convergence at the continental margin is the absence of a bathymetric trench

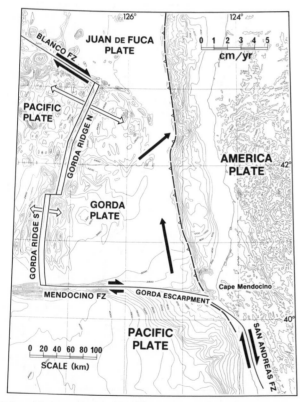

Fig. 10. Features and movements of the Explorer plate and northern triple junction region. Arrows represent motion vectors in cms/yr calculated from magnetic anomalies. Split arrows are transform/strike slip motion, half the motion on each side. Solid arrows at the margin are motion relative to continent and N. America plate. Open arrows are ridge spreading, half the motion on each side of the ridge.

along the base of the continental slope.
However, reflection seismic profiling (e.g.
Barr, 1974; Chase et al., 1976; McManus et al.,
1971) and gravity interpretations (Srivastava,
1974) show that the ocean crust does dip towards
the margin at 5° or greater but that the high
sediment rates of the middle and late Pleistocene
may have effectively filled any topographic
depression along the base of the slope (Scholl,
1974).

The immediate effects of subducting oceanic
crust at the continental edge should be evidence
of compression and underthrusting in the
sedimentary structures of the slope. A number of
detailed studies along the continental margin off
Vancouver Island (Tiffin et al., 1972), Washington
(Silver, 1972; Barnard, 1978), Oregon (Kulm and
Fowler, 1974) and California (Silver, 1971) have
shown compression, imbricate thrusting and uplift
in late Tertiary and younger sediments. A
characteristic anticlinal ridge at the base of
the continental slope terrace (Fig. 11) has been
interpreted (Carson et al., 1974; Kulm and
von Heune, 1973) as the outer front of the
compression and used to confirm that convergence
rates, at least off Washington and Oregon, are of
the order of those estimated from magnetic
anomalies. Despite this evidence, seismicity in
the convergence zone of the continental slope is
minimal and no events have been unequivocally
shown to be associated with the underthrusting
process during the 80 years that seismic events
have been recorded.

The Arc-Trench Gap

In the area of the continent between the coast
and the Cascade volcanic chain, general
seismicity with event magnitudes up to 7.3 has
been observed from Vancouver Island to Puget
Sound (Milne et al., 1978; Crosson, 1972) and in
northern California (e.g. Chandra, 1974). The
majority of events are located within crustal
depths (Crosson, 1976) but some deeper and larger
events have been recorded. The lack of a suite
of deep hypocentres has for many years been
advanced as critical evidence against the
existence of a subducting slab of oceanic
lithosphere beneath the region. In two areas
however, detailed location of hypocentres has
confirmed the existence of a dipping structure
beneath the crust with a characteristic aseismic
zone immediately above it (Smith and Knapp, 1979;
Crosson, 1976). The absence of events deeper
than 70 km has been ascribed to the rapid
re-heating of the descending slab and shown to be
consistent with relations between earthquake
depths and rates of subduction at other active
margins (Riddihough and Hyndman, 1978).

Fault plane solutions calculated for crustal
events in the region (Figure 5) predominantly
indicate either NW-SE right-lateral strike-slip
or NE-SW left-lateral strike-slip, both
solutions being the theoretical result of north-

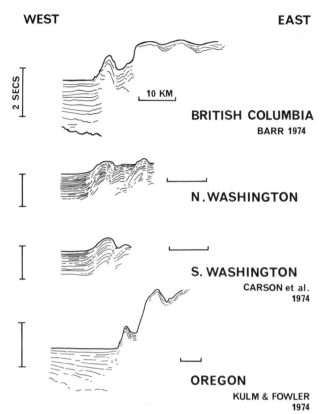

WEST   EAST

2 SECS

10 KM

BRITISH COLUMBIA
BARR 1974

N.WASHINGTON

S. WASHINGTON
CARSON et al.
1974

OREGON
KULM & FOWLER
1974

Fig. 11. Line drawings from reflection seismic
profiles across the foot of the continental
slope along the Juan de Fuca plate margin
showing compressional anticlinal structure.

south compression (Chandra, 1974; Crosson, 1972;
Milne et al., 1978; Rogers, 1979). The NW-SE
solution has generally been preferred but except
in northern Vancouver Island (Rogers, 1979), the
origin of a regional north-south compression
remains unexplained. Solutions for some deeper
events in the Puget Sound area which suggest
eastward dipping normal faulting have been
assigned to the downgoing slab (McKenzie and
Julian, 1971; Chandra, 1974).

The lack of earthquakes in any part of the
margin with a definite thrust solution has been
a more difficult feature to explain. At
present, the alternatives of essentially
aseismic subduction associated in some way with
the slow, oblique approach of the Juan de Fuca
plate to the margin, or of major thrust earthquakes with a repeat period of many hundreds of
years which have not yet been observed, seem to
offer the most plausible solutions.

As observed across other active margins, a
zone of low heat-flow in the arc-trench gap,
backed by a marked increase near the volcanic
arc, has been observed from a series of
measurements along the margin (Hyndman, 1976)
(Fig. 12). The effect is assumed to be the

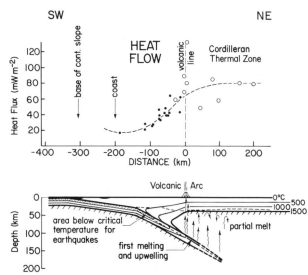

Fig. 12. Heat flow profile across the margin of British Columbia with theoretical temperature cross-section (from Keen and Hyndman, 1979).

result of heat absorbtion and endothermic phase changes in the descending oceanic material. The time scale of heat transmission through the crust would confirm that subduction occurred as recently as 4 or 5 million years ago.

A number of studies of geomagnetic variations in British Columbia (Caner et al., 1971) have confirmed the existence of a broad area of shallow, high conductivity east of the volcanic chain. The complexities of coastal and island effects (e.g. Nienaber et al., 1979) have so far prevented the clear determination of an effect which can be ascribed to a downgoing slab, however a geomagnetic anomaly along the volcanic chain in Washington has been suggested.

Coverage of gravity measurements over both the land and sea areas of the Juan de Fuca plate is extensive and a number of interpretations of the structure of the margin have been made (e.g. Dehlinger et al., 1971). Regional anomaly patterns are broadly parallel to the margin and the volcanic arc. Narrow linear negative anomalies occur along the continental slope and shelf and positive Bouguer anomalies over Vancouver Island and the Washington-Oregon Coast Range. A strong regional negative anomaly begins at the volcanic front (Fig. 13). Since 1970 and the application of a plate tectonic model to the region, there have been few re-assessments of the gravity interpretations (Srivastava, 1973; Stacey, 1973); however, it would seem that the gravity field is compatible with a subduction model (Riddihough, 1979). One problem is the discrepancy between gravity and seismically determined crustal thicknesses under Vancouver Island. This was discussed at length by Stacey (1973) and a solution involving the development of high-density - low-velocity

metamorphic material in the wedge overlying the downgoing slab has been proposed (Riddihough, 1979) (Fig. 13).

Seismic refraction experiments in the Pacific NW region have not been specifically designed to detect the existence of a dipping slab but in general have not provided conclusive evidence against it. Crustal thickness in the arc-trench gap is generally 20 km or less in the area of the Washington-Oregon coast range (Berg et al., 1966). A more recent refraction experiment shows that it thickens eastwards beneath the Cascade volcanic chain to near 30 km (Johnson and Couch, 1970). A detailed profile across the British Columbia section of the margin suggests that there is a discontinuity between Vancouver Island and the mainland (Berry and Forsyth, 1974) and confirms the anomalous structure beneath Vancouver Island of crustal velocities down to depths of near 50 km.

Recent work on the geology and palaeomagnetic history of the arc-trench region, particularly in Washington and Oregon, has established that much of the crust south of Vancouver Island is essentially oceanic and may represent a sea-mount province which has moved in with the converging oceanic plate and 'choked' the subduction zone. The shift of the subduction zone from the eastern to the western side of the Coast Range block (Dickinson, 1970; Muller, 1977) apparently took place since 40 m.y.b.p. Clockwise rotations of crustal blocks of material either during or after this 'choking' process are now well established (Simpson and Cox, 1977; Beck and Burr, 1979) and imply rotations of up to 2° per m.y. until at least Miocene time. These clockwise motions are not yet clearly explained but seem to be compatible with the general 'right-lateral' type of plate interaction produced by the north-easterly convergence of the Juan de Fuca plate system along a principally north-south continental margin. They are similar to the observed rotations in the offshore plates referred to earlier, and may imply periodic coupling of parts of the continent to the converging plate system. Various mechanisms for microplate rotations along the western margin of N. America have been discussed by Beck (1976), Menard (1978) and Simpson and Cox (1977).

The Volcanic Arc

The Quaternary and late Tertiary volcanic arc facing the Juan de Fuca plates stretches from Meager Mt. in British Columbia to Lassen Peak in California and has been studied in considerable detail. (Fig. 14). It has passed through a number of surges and declines but continues to be active into the present. Since 1800, seven of the volcanoes have been active, the most recent major eruptions being Lassen Peak (1914 - 17) and Mt. St. Helens (1980).

The relationship between the volcanism of the

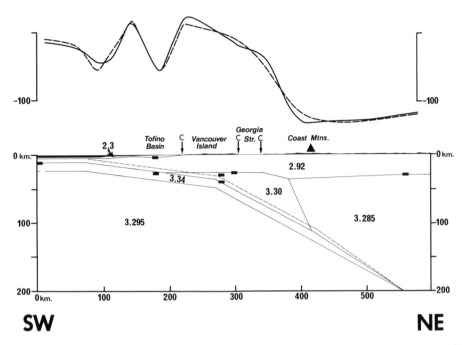

Fig. 13. Gravity and structural interpretation across the margin of British Columbia (from Riddihough, 1979). Blocks are seismic control points, C marks the coastline. Numbers are densities in gms/cm^3. Solid lines mark density boundaries, dashed line is upper surface of oceanic crust. In gravity profile, solid line is observed gravity, dashed is calculated.

chain and the dynamics of the Juan de Fuca plate system is far from clear. Given that convergence rates suggest that there is likely to be a lag of around 5 m.y. between material being subducted at the trench and any resultant volcanism, the north-south extent of the Quaternary arc seems to fit the geometry of recent subduction satisfactorily (Souther, 1977; Snyder et al., 1976). However, although the products of the arc are calc-alkaline and typical of Benioff zone magmatism (Souther, 1977), the proportion of andesitic to basaltic material seems to have steadily decreased over the last 40 m.y. Analyses of the composition and trace elements of the rocks, at least in the Cascades, also point to a primary mantle origin rather than an origin by melting of subducted material (McBirney, 1978). Palaeomagnetic evidence (Beck and Burr, 1979) suggests that the older volcanic rocks underlying the late Tertiary chain may have participated in the rotations discussed in the preceding section.

Surges and quiet periods in the level of volcanism along the arc have not so far been satisfactorily related to changes in the dynamics of the convergence process. This is epitomised by the lack of any demonstrated connection between the rapid development of the large Pleistocene High Cascade andesitic cones in the last 2 million years (McBirney, 1978) and details of the convergence history along the margin as modelled by Riddihough (1977).

Although the relation has been discussed both in the geometric (e.g. Stacey, 1973) and petrogenetic sense (e.g. Church, 1976), it is clearly one in need of much closer examination and evaluation.

Summary Comments

The work carried out in the Juan de Fuca plate region during the period of the Geodynamics Project has ensured that many basic geophysical and geological parameters are now known in considerable detail. What is remarkable is that despite this, a clear understanding of the geodynamics and driving forces of the system still evades us. The area is evidently complex but is nevertheless of global importance because in many cases it may provide limiting cases to some of the basic principles of the plate tectonic model. Examples of this are the minimum size of a rigid oceanic plate, the minimum age of an oceanic plate that can still subduct beneath a continental margin and the minimum depth for an active Benioff zone. It also provides well studied examples of the relationship between oblique subduction and seismicity, between oblique subduction and block rotations, and between subduction and volcanism. At its northern and southern ends it provides highly detailed examples of the contemporary evolution of triple junctions.

The acceptance that the Juan de Fuca plate

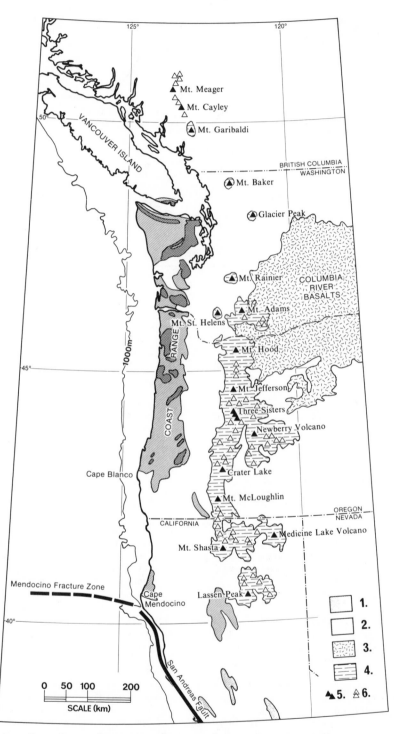

Fig. 14. Sketch map of geology of arc-trench zone of Juan de Fuca plate. 1, Tertiary marine deposits; 2, Tertiary pillow basalts; 3, Columbia River, middle Tertiary plateau basalts; 4, Quaternary terrestrial volcanics; 5, major Quaternary volcanoes; 6, secondary Quaternary volcanic centres.

system has moved over the last few million years and continues to do so today, is a critical step on the way to the acceptance that the geodynamic processes seen in this region, however complex, are applicable elsewhere on the globe. The last 9 years has seen the changeover from questioning the very existence of the system to the investigation of the details of its kinematics. With the amount of information currently available, we are hopeful that the next decade may provide answers to fundamental questions concerning its dynamics and energy balance.

Acknowledgements. This is the final report of the Juan de Fuca Plate Study Group to Working Group 2 of the Inter-Union Commission on Geodynamics. It is Contribution of the Earth Physics Branch No. 811.

Bibliography

The following bibliographies are not complete compilations of all geological and geophysical work in the region. They are selected to provide the appropriate background for an understanding of the geodynamics of the region and to provide key entry points into the accumulated literature.

Atwater, T., Implications of plate tectonics for the Cenozoic tectonic evolution of western North America, Geol. Soc. Amer. Bull. 81, 3513-3536, 1970.

Atwater, T. and J.D. Mudie, Detailed near-bottom geophysical study of the Gorda Rise, J. Geophys. Res., 78, 8665-8686, 1973.

Barnard, W.D., The Washington continental slope: Quaternary tectonics and sedimentation, Marine Geol., 27, 79-114, 1978.

Barr, S.M., Structure and tectonics of the continental slope west of Vancouver Island, Can. J. Earth Sci., 11, 1187-1199, 1974.

Barr, S.M., Seamount chains formed near the crest of the Juan de Fuca Ridge, north-east Pacific Ocean, Marine Geol., 17, 1-19, 1974.

Barr, S.M. and R.L. Chase, Geology of the northern end of Juan de Fuca Ridge and sea-floor spreading, Can. J. Earth Sci., 11, 1384-1406, 1974.

Beck, M.E., Jr., Discordant paleomagnetic pole positions as evidence of regional shear in the western Cordillera of North America, Amer. J. Sci., 276, 694-712, 1976.

Beck, M.E., Jr., and C.D. Burr. Paleomagnetism and tectonic significance of the Goble volcanics of southwestern Washington, Geology, 7, 175-179, 1979.

Berg, J.W., Trembly, L., Emilia, D.A., Hutt, J.R., King, J.M., Long, L.T., McKnight, W.R., Sarmah, S.K., Souders, R., Thiruvathukal, J.V., and D.A. Vossler, Crustal refraction profile, Oregon Coast Range, Bull. Seism. Soc. Amer., 56, 1357-1362, 1966.

Berry, M.J., and D.A. Forsyth, Structure of the Canadian Cordillera from seismic refraction and other data, Can. J. Earth Sci., 21, 182-208, 1975.

Bevier, M.L., Armstrong, R.L., and J.G. Souther, Miocene peralkaline volcanism in west central British Columbia: its temporal and plate tectonic setting, Geology, 7, 389-392, 1979.

Caner, B., Auld, D.R., Dragert, H., and P.A. Camfield, Geomagnetic depth sounding and crustal structure in Western Canada, J. Geophys. Res., 76, 7181-7201, 1971.

Carson, B.J., Yuan, P.B., Myers, P.B. and W.B. Barnard, Initial deep-sea sediment deformation at the base of the Washington continental slope: a response to subduction, Geology, 3, 561-564, 1974.

Chandra, U., Seismicity, earthquake mechanisms and tectonics along the western coast of North America from 42° N to 61° N, Bull. Seism. Soc. Amer., 64, 1529-1549, 1974.

Chase, R.L., J. Tuzo Wilson Knolls: Canadian Hot Spot, Nature, 266, 344-346, 1977.

Chase, R.L., Tiffin, D.L., and J.W. Murray, The western Canadian Continental Margin, Can. Soc. Petrol. Geol. Memoir 4, 701-722, 1975.

Church, S.E., The Cascade Mountains revisited: a re-evaluation in light of new lead isotopic data, Earth Planet. Sci. Lett., 29, 175-188, 1976.

Couch, R., Seismicity and crustal structure near the north end of the San Andreas Fault system, in 'San Andreas Fault in northern California', Calif. Div. Mines and Geology Special Report, 1979.

Crosson, R.S., Small earthquakes, structure and tectonics of the Puget Sound region, Bull. Seism. Soc. Amer., 1133-1171, 1972.

Crosson, R.S., Crustal structure modelling of earthquake data 2. Velocity structure of the Puget Sound Region, Washington, J. Geophys. Res., 81, 3047-3054, 1976.

Davis, E.E. and C.R.B. Lister, Tectonic structures on the Juan de Fuca Ridge, Geol. Soc. Amer. Bull., 88, 346-363, 1977.

Davis, E.E., Lister, C.R.B. and B.T.R. Lewis, Seismic structure on Juan de Fuca Ridge: ocean bottom seismometer results from the median valley, J. Geophys. Res., 81, 3541-3555, 1976.

Davis, E.E., Lister, C.R.B., Wade, U.S. and R. D. Hyndman, Detailed heat flow measurements over the Juan de Fuca ridge system and implications for the evolution of hydrothermal circulation in young ocean crust, J. Geophys. Res., 85, 299-310, 1980.

Dehlinger, P., R.W. Couch, D.A. McManus and M. Gemperle, Northeast Pacific Structure in "The Sea", Ed. A.E. Maxwell, 14, Pt. 2, 133-189, 1971.

Dickinson, W.R., Sedimentary basins developed during evolution of Mesozoic-Cenozoic arc-trench system in North America, Can. J. Earth Sci., 13, 1268-1287, 1976.

Elvers, D., Srivastava, S.P., Potter, K., Morley, J., and D. Sdidel, Asymmetric spreading across the Juan de Fuca and Gorda Rises as obtained from a detailed magnetic survey, Earth Planet. Sci. Letts., 20, 211-219, 1973.

Emilia, D.A., Berg, J.W., and W.E. Bales, Magnetic anomalies off the northwest coast of the United States, Geol. Soc. Amer. Bull., 79, 1053-1062, 1968.

Herd, D.G., Intracontinental plate boundary east of Cape Mendocino, California, Geology, 6, 721-725, 1978.

Hyndman, R.D., Heat flow measurements in the inlets of southwestern British Columbia, J. Geophys. Res., 81, 337-349, 1976.

Hyndman, R.D., Riddihough, R.P. and R. Herzer, The Nootka Fault Zone - a new plate boundary off Western Canada, Geophys. J. Roy. Astr. Soc., 58, 667-683, 1979.

Hyndman, R.D., Rogers, G.C., Bone, M.N., Lister, C.R.B., Wade, U.S., Barrett, D.L., Davis, E.E., Lewis, T., Lynch, S., and D. Seeman, Geophysical measurements in the region of the Explorer ridge off western Canada, Can. J. Earth Sci., 15, 1508-1525, 1978.

Johnson, S.H. and R.W. Couch, Crustal structure in the North Cascade Mountains of Washington and British Columbia from seismic refraction measurements, Bull. Seism. Soc. Amer., 60, 1259-1269, 1970.

Johnson, S.H. and P.R. Jones, Microearthquakes located on the Blanco Fracture zone with sonobuoy arrays, J. Geophys. Res., 83, 255-261, 1978.

Jones, P.R. and S.H. Johnson, Sonobuoy array measurements of active faulting on the Gorda ridge, J. Geophys. Res., 83, 3435-3440, 1978.

Keen, C.E. and R.D. Hyndman, Geophysical review of the continental margins of eastern and western Canada, Can. J. Earth Sci., 16, 712-747, 1979.

Klitgord, K.D., Huestis, S.P., Mudie, J.D. and R.L. Parker, An analysis of near bottom magnetic anomalies; sea floor spreading and the magnetized layer, Geophys. J. Roy. Astr. Soc., 43, 387-424, 1975.

Korgen, B.J., Badvarrson, G., and R.S. Mesecan, Heat flow through the floor of the Cascadia Basin, J. Geophys. Res., 76, 4758-4774, 1971.

Kulm, L.D., and G. Fowler, Oregon continental margin structure and stratigraphy: A test of the imbricate thrust model, in: The Geology of Continental Margins, Edited by: C.A. Burk and C.L. Drake, Springer-Verlag, New York, 261-283, 1974.

Larson, R.L., Bathymetry, magnetic anomalies and plate tectonic history of the mouth of the Gulf of California, Geol. Soc. Amer. Bull., 83, 3345-3360, 1972.

McBirney, A.R., Volcanic evolution of the Cascade Range, Ann. Rev. Earth Planet. Sci., 6, 437-456, 1978.

McLeod, N.S., Tiffin, D.L., Snavely, P.D., and R.G. Currie, Geologic interpretation of magnetic and gravity anomalies in the Strait of Juan de Fuca, U.S. - Canada, Can. J. Earth Sci., 14, 223-238, 1977.

McManus, D.A., Holmes, M.L., Carson, B. and S.M. Barr, Late Quaternary tectonics, northern end of Juan de Fuca Ridge, Marine Geol., 12, 141-164, 1972.

Malacek, S.J., and R.M. Clowes, Crustal structure near Explorer Ridge from a marine deep seismic sounding survey, J. Geophys. Res., 83, 5899-5912, 1978.

Menard, H.W., Fragmentation of the Farallon Plate by pivoting subduction, J. Geol., 86, 99-110, 1978.

Milne, W.G., Rogers, G.C., Riddihough, R.P., Hyndman, R.D. and G.A. McMechan, Seismicity of western Canada, Can. J. Earth Sci., 15, 1170-1193, 1978.

Muller, J.E., Evolution of the Pacific margin, Vancouver Island, and adjacent regions, Can. J. Earth Sci., 14, 2062-2085, 1977.

Nienaber, W., Dosso, H.W., Law, L.K., Jones, F.W. and V. Ramaswamy, An analogue model study of electromagnetic induction in the Vancouver Island region, J. Geomag. Geoelectr., 31, 115-132, 1979.

Potter, K., Morley, J. and D. Elvers, Magnetics Maps 12042-12, 13242-12, NOS Sea Map Series, N. Pacific Ocean, U.S. Dept. of Commerce, NOAA, Scale 1:1,000,000, 1974.

Raff, A.D. and R.G. Mason, Magnetic survey off the west coast of North America, 40°N to 52°N latitude, Geol. Soc. Amer. Bull., 72, 1267-1270, 1961.

Riddihough, R.P., A model for recent plate interactions off Canada's west coast, Can. J. Earth Sci., 14, 384-396, 1977.

Riddihough, R.P., The Juan de Fuca Plate. Trans. Amer. Geophys. Un. (EOS), 59, 836-842, 1978.

Riddihough, R.P., Structure and gravity of an active margin - British Columbia and Washington, Can. J. Earth Sci., 16, 350-363, 1979.

Riddihough, R.P. and R.D. Hyndman, Canada's active western margin: the case for subduction, GeoScience Canada, 3, 269-278, 1976.

Rogers, G.C., Earthquake fault plane solutions near Vancouver Island. Can. J. Earth. Sci., 16, 523-531, 1979.

Rogers, G.C., Juan de Fuca Plate Map - Fault Plane solutions. Earth Physics Branch, Ottawa, Open File 80-4, 1980.

Scholl, D.A., Sedimentary sequences in North Pacific trenches. in: The Geology of Continental Margin, eds. C.A. Burk and C.L. Drake, Springer-Verlag, 493-504, 1974.

Seeber, L., Barazangi, M. and N. Nowroozi, Microearthquakes, seismicity and tectonics of coastal northern California, Bull. Seism. Soc. Amer., 60, 1669-1699, 1970.

Seely, D.R., The significance of landward vergence and oblique structural trends of trench inner slopes, in: Island Arcs, Deep Sea Trenches and Back Arc Basins, Eds. M. Talwani and W.C. Pitman, Amer. Geophys. Union., 187-198, 1977.

Silver, E.A., Small plate tectonics in the northeastern Pacific, Geol. Soc. Amer. Bull., 82, 3491-3496, 1971a.

Silver, E.A., Transitional tectonics and late Cenozoic structure of the continental margin of northernmost California, Geol. Soc. Amer. Bull., 82, 1-22, 1971b.

Silver, E.A., Tectonics of the Mendocino Triple Junction, Geol. Soc. Amer. Bull., 82, 2965-2978, 1971c.

Silver, E.A., Pleistocene tectonic accretion of the continental slope off Washington, Marine Geol., 13, 239-249, 1972.

Silver, E.A., von Heune, R. and J.K. Crouch, Tectonic significance of the Kodiak-Bowie seamount chain, NE Pacific, Geology, 2, 147-150, 1974.

Simpson, R.W. and A. Cox, Palaeomagnetic evidence for tectonic rotation of the Oregon Coast Range, Geology, 5, 585-589, 1977.

Smith, R.B., Shuey, R.T., Pelton, J.R., and J.P. Bailey, Yellowstone hotspot: Contemporary tectonics and crustal properties from earthquake and aeromagnetic data, J. Geophys. Res., 82, 3665-3676, 1977.

Smith, S.W. and J.S. Knapp, The northern termination of the San Andreas Fault, in 'San Andreas fault in northern California', Calif. Div. of Mines & Geology Special Report, 1979.

Souther, J.G., Volcanism and Tectonic environments in the Canadian Cordillera - a second look, in: Volcanic Regimes in Canada, edited by W.R.A. Baragar, L.C. Coleman, J.M. Hall, Geol. Ass. Can. Spec. Pap., 16, 3-24, 1977.

Snyder, W.S., Dickinson, W.R. and M.L. Siberman, Tectonic implications of space-time patterns of Cenozoic magmatism in the western United States, Earth Plan. Sci. Letts., 32, 91-106, 1976.

Srivastava, S.P., Interpretation of gravity and magnetic measurements across the continental margin of British Columbia, Canada, Can. J. Earth Sci., 10, 1664-1677, 1973.

Wilson, J.T., Transform faults, ocean ridges and magnetic anomalies southwest of Vancouver Island, Science, 150, 482-485, 1965.

Srivastava, S.P., Barrett, D.L., Keen, C.E., Manchester, K.S., Shih, K.G., Tiffin, D.L., Chase, R.L., Thomlinson, A.G., Davis, E.E., and C.R.B. Lister, Preliminary analysis of geophysical measurements north of Juan de Fuca Ridge, Can. J. Earth Sci., 8, 1265-1281, 1971.

Stacey, R.A., Gravity anomalies, crustal structure and plate tectonics in the Canadian Cordillera, Can. J. Earth Sci., 10, 615-628, 1973.

Stacey, R.A., Plate tectonics, volcanism and the lithosphere in British Columbia, Nature, 250, 133-134, 1974.

Tatham, R.H. and J.M. Savino, Faulting mechanisms for two oceanic earthquake swarms, J. Geophys. Res., 73, 3821-3846, 1974.

Tiffin, D.L., Cameron, B.E.B., and J.W. Murray, Tectonic and depositional history of the continental margin off Vancouver Island, B.C., Can. J. Earth Sci., 9, 280-296, 1972.

Tiffin, D.L., and R.P. Riddihough, Gravity and magnetic survey off Vancouver Island, 1975, Report of Activities, Part A, Geological Survey of Canada, Paper 77-1A, 1977.

Tobin, P.G., and L.R. Sykes, Seismicity and tectonics of the northeast Pacific Ocean, J. Geophys. Res., 72, 3821-3846, 1968.

Vine, F.J. and J.T. Wilson, Magnetic anomalies over a young oceanic ridge off Vancouver Island, British Columbia, Science, 150, 485-489, 1965.

Vogt, P.R. and G.L. Johnson, Transform faults and longitudinal flow below the mid-oceanic ridge, J. Geophys. Res., 80, 1399-1428, 1975.

Vogt, P.R. and G.R. Byerly, Magnetic anomalies and basalt composition in the Juan de Fuca - Gorda Ridge area, Earth Plan. Sci. Letts., 33, 185-207, 1976.

Von Heune, R., and L.D. Kulm, Tectonic summary of Leg 18, in: Initial Reports of the Deep Sea Drilling Project XVIII, Kulm, L.D., von Heune, R. et al., U.S. Government Printing Office, Washington, D.C., 961-976, 1973.

Wetmiller, R.J., An earthquake swarm on the Queen Charlotte Island fracture zone, Bull. Seism. Soc. Amer., 61, 1489-1505, 1971.

BIBLIOGRAPHY CONCERNING MEXICAN GEODYNAMICS

Luis Del Castillo (Complemented by R. Cabre, Editor)

Uranio Mexicano

Introduction

The bibliography introduced here deals with geodynamic problems of Mexico and neighbouring seas, but it does not purport to be complete: while several publications of wide distribution are included, emphasis is placed on Mexican publications presenting research achieved during the Geodynamics Project.

Each year the Unión Geofísica Mexicana (Instituto de Geofísica, UNAM-Ciudad Universitaria-México 20, D.F. Mexico) has a meeting where most of the geophysical investigations carried on in Mexico are presented, many of high geodynamical interest especially in the section devoted exresssly to geodynamics; the abstracts are published under the title Unión Geofísica Mexicana - Reunión Anual (19XX) and will not be referenced separately here.

Geophysical, geological and geomorphological investigators have contributed to a better knowledge of plate boundaries in southeastern Mexico and their movements, especially along the Polochic-Motagua system of regional faults close to Guatemala border and in the Gulf of California. The most relevant feature there is the recent right lateral fault movement (Moore, 1973; Alvarez and Del Río, 1975; Del Castillo, 1975, 1976, 1977; Lomnitz, 1975; López Ramos, 1975; 1976; Lugo, 1976, Reyes et al., 1976; Flores and González, 1976; Schwartz, 1976; Reid, 1976; González and Flores, 1978; Sandoval et al. 1978).

Gravity surveies have been expanded in areas of high geodynamical interest, such as subductions zones along the Pacific Coast, the Trans-Mexican Volcanic Belt and the Gulf Coast Plain (Monges et al., 1977). The Bouguer anomaly decreases from the Pacific towards the Mexican continent (Del Castillo, 1977) suggesting different subduction dips of the Cocos plate. This is supported by differential seismic activity towards the coastal zone of the Gulf of Mexico (Moore and Del Castillo, 1974; Del Castillo and Urrutia, 1975).

Magnetic surveys together with gravity measurements in the Gulf of Mexico and in the Pacific have become of economical interest since they yield interesting results concerning mineralization, geothermal possibilities and hydrocarbons (Bayer, 1976; González, 1976, 1977; Mendive, 1976

Del Castillo, 1977; Flores et al., 1976; Mejía, 1977, Navarro, 1977; Salas, 1977).

Seismic refraction profiles, the Oazaca and Acapulco projects, were carried on as a collaboration of several Mexican institutions with the Universities of California, Texas and Wisconsin (Del Castillo and Lomnitz, 1975). Previous geological estimations about that tectonically active region were confirmed and completed. Seismic velocities across the rocks and structures of the Middle America Trench are quite different from those sited in the Mexican high plateau. Focal mechanisms have contributed to the study of internal dynamics (Jiménez, 1977; Hanus and Vanek, 1977-78).

Detailed gravity studies controlled by seismic profiling confirmed crustal thickness is of 19 ± 1 km in the central part of the Gulf of Mexico, 45 km in the Zacatecas State and of more than 25 km in the Yucatan peninsula.

Heat flow measurements (Blackwell, 1976; García, 1976) and different local data from the southeastern part of the country suggest dispersion centers (Del Castillo, 1976) associated with earthquake swarms (Figueroa, 1976; Rodríguez et al., 1976) originating in Cocos and Caribbean plate movements.

Seismic information was improved after installation of a telemetered seismic network, called RESMAC (Gil and Lomnitz, 1976; Garza and Lomnitz, 1976). Valuable assistance came from the tectonic and geomorphologic interpretation of computer processed ERTS and SKYLAB images (Salas, 1976; Guerra, 1977; Meritano, 1977; Guzmán, 1977).

Combined geomorphology, geology and geophysics studies have been applied mostly to find relations between faulted and mineralized areas of economic interest such as the Mexico Basin and the Coahuila Peninsula.

Volcanological studies have assisted inproviding better knowledge of several subjects (Mooser, 1972; Wood, 1974).

Paleomagnetic measurements suggest a rotation of Mexico (Nairn et al., 1975; Urrutia and Pal, 1976, 1977; Pal and Urrutia, 1977; Urrutia, 1978).

References

Alvarez, R. and L. Del Río, Transcontinental faulting: evidence for superimposed subduction in Mexico? (abstract), Trans.Am.Geoph.Union, 56, p. 1066, 1975.

Comínquez, A.H., J.H. Sandoval and L. Del Castillo Aporte gravimétrico en el análisis tectonofísico del Golfo de México, Rev.Asoc.Mexic.Geofís. Explor. 1977.

Dean, B.W. and C.L. Drake, Focal mechanism solutions and tectonics of the Middle America arc, J. Geol. 86, p. 111-128, 1978.

Del Castillo, L. et al., Geophysical investigation in the Cocos Plate: an active plate, 13th Pacific Science Congress (abstract), p. 271, 1975.

Del Castillo, L.,The Caribbean problem: a complex fault system, Proceedings II - CICAR Symposium, 1976.

Del Castillo, L., Cocos Plate Study Group, In "Geodynamics: Progress and Prospect", Am.Geoph. Union, edited by C.L. Drake, p. 20-22, 1976.

Del Castillo, L., La contribución de la Geodinámica al conocimiento de la tectónica del subsuelo y del fondo marino, Memoria III Congreso de la Academia Nacional de Ingeniería, A. C., p. 56-61, 1977.

Del Castillo, L., Centros de Dispersión asociados con Geotermia en México, Memoria I Congreso Ibero-Latinoamericano de Geofísica, 1977.

Del Castillo, L., Implicación Eje Volcánico-Placa de Cocos-Placa del Caribe, Memoria V Congreso Venezolano de Geología, 1977.

Del Castillo, L., Geodynamic evolution of Coco's-Caribbean plates in the eastern Mexican boundary, 8th Caribbean Geological Conference, p. 37, 1977.

Del Castillo, L. and C. Lomnitz, El experimento Oaxaca, Noticiario Inst. geof., UNAM. 6, p. 5-8, 1975.

Flores, C. and G. Gonzalez, Un estudio geofísico en la Boca del Golfo de California, tesis profesional, Facultad de Ingeniería, UNAM. 1976.

García, M. Radiactividad y calor en la depresión central del Estado de Chiapas (Chiapa de Corzo), B.SC. Thesis, I.P.N., p. 95, 1976.

Gil, J. and C. Lomnitz, El sistema Resmac y la exploración de los océanos, Memoria I Reunión Anual sobrd la zona Económica exclusiva, 1976.

González, T., Identificación de estructuras geológicas utilizando transformaciones conformes, Memoria III Congreso de la Academia de Ingeniería, C.A., p. 70-73, 1977.

González, J.J. and C.F. Flores, Evidencia sísmica de refraccion de una zona de esparicimiento en la boca del Golfo de California, Anales del Inst. de Geofís., UNAM 22-23, p. 117-129, 1978.

Guerra Peña, Interpretación de la Tectónia Mexicana en las imágenes del satélite artificial ERTS, Comisión de Estudios del Territorio Nacional, p. 19, 1977.

Guzmán, A. Proyecto de Percepción Remota aplicada (Automatización de imágenes), Inst. Inv.Mat. Aplicadas, 1977.

Hanus, V. and J. Vanek, Subduction in the Cocos plate and Deep Active Fracture Zones of Mexico, Geofis. Intern., 17, p. 14-53, 1977-78.

Jiménez, Z. Mecanismo focal de siete temblores (m_b 5.5) ocurridos en la región de Orizaba,

México, en el período de 1928 a 1973, Tesis profesional, Facultad de Ciencias, UNAM, México, 1977.

Lomnitz, C., Global tectonics and seismic risk, Elsevier Publ. 1975.

Lomnitz, C. and J. Gil. Resmac: the new Mexican seismic array, EOS 57, p. 68-69, 1976.

Lomnitz, C.A. procedure for eliminating the indeterminacy in focal depth determinations, Bul.Seism.Soc.Am. 67, p. 533-536, 1977.

López, E. Geología general de México, Edit.México, 4th edit. 1975.

Lozano, F. Evaluación petrolífera de la península de Baja California, México, Asoc.Mex.Geol. Petr. A.C. XXVII, p. 420.

Lugo, J. De la Geomorfología clásica a la geomorfología estructural, Ciencia y Desarrollo, 8, p. 27-32, 1976.

Mejía, O. Exploración petrolera en el período 1977-1982, México, Gaceta Soc.Geol.Mex. III, p. 10-12, 1977.

Mendive, L. et al., Métodos matemáticos y algoritmos en estudios de Geofísica Marina aplicados en México, Memoria I Reunión Nal. sobre la Zona Económica Exclusiva, 1976.

Meritano, J. Estudio tectónico preliminar del Istmo de Tehuantepec con base en imágenes del Satélite ERTS, Ciencia y Desarrollo, 16, p. 51-52, 1977.

Monges, J. et al., Reporte Nacional de Geofísica para el IPGH, p. 34, 1977.

Moore, D.G., Plate-edge deformation and crustal grouth, Gulf of California, tesis profesional, Facultad de Ingeniería, UNAM, 1976.

Moore, G.W. and L. Del Castillo, Tectonic evolution of the southern Gulf of Mexico, Bull.Geol. Soc.Am. 85, p. 607-618, 1974.

Mooser, F., The Mexican volcanic belt structure and tectonics, Geofís. Intern. 12, p. 55-70, 1972.

Nairn, A.E.M., J.F.W. Negendank, H.C. Noltimier and T.J. Schmitt, Paleomagnetic investigations of the Tertiary and Quaternary igneous rocks. X: The ignimbrites and lava units west of Durango, Mexico, Neues Jahrb.Geol.Paleont., Monatsh. 11, p. 664-678, 1975.

Pal, S. and J. Urrutia, Paleomagnetism, geochronology and geochemistry of some igneous rocks from Mexico and their tectonic implications, Proc. IV International Gondwana Symp., Calcutta, India, 1977.

Reichle, M.A. seismological study on the Gulf of California: sonobuoy and teleseismic observations and tectonic implications, Ph.D. Thesis, University of California, San Diego, 1975.

Reid, I, The Rivera Plate: a study in seismology and plate tectonics, Ph.D. dissertation, University of California, San Diego, 1976.

Salas, P. Carta y Provincias metalogenéticas de la República Mexicana, Consejo de Recursos Minerales, publ. 21-E, p. 227, 1976.

Salas, P. El potencial minero de México, Gaceta Soc. Geol. Mex. III, p. 6-8, 1977.

Sandoval, H., A.H., Comínguez and L. Del Castillo, Modelo geodinámico de la estructura del Golfo de México, Anales Inst. Geofís., 1976-1977, 22-23, p. 153-165, 1978.

Schwartz, P. Active faulting along the Caribbean-North America plate Boundary in Guatemala, V Reunión de Geólogos de América Central (abstract) 1977.

Urrutia, J. Paleointensidad del campo geomágnético determinada de rocas ígneas de México, Anales Inst.Geofís. 22-23, p. 211-216, 1978.

Urrutia, J. and S. Pal, Further paleomagnetic evidence for a possible Tectonic Rotation of Mexico, Geofís. Intern. 16, p. 255-259, 1976.

Urrutia, J. and S. Pal, Paleomagnetic data from Tertiary igneous rocks, NE Jalisco, Mexico, Earth Planet.Sci.Lett.36, p. 202-206, 1977.

Wood, C.A. Reconnaissance Gephysics and Geology of the Pinacate Craters, Sonora, Mexico, Bull. Volc.38, p. 149-172, 1974.

SUMMARY AND BIBLIOGRAPHY OF CARIBBEAN
GEODYNAMIC INVESTIGATORS

J.F. Lewis, George Washington University
W. MacDonald, State University of
New York, Binghamton
A.L. Smith, University of Puerto Rico
J.D. Weaver, (Deceased) University of
Puerto Rico

Editor's Note:

John Weaver was a member of working group 2 from its inception through his untimely death late in 1979. He was also the reporter for the Caribbean region for the U.S. Geodynamics Committee. John and his colleagues prepared a summary of geodynamics related activities in the Caribbean prior to his death and, in addition, assembled a bibliography of investigators in the Caribbean during the 1970's. The summary is presented here and the bibliography is included on microfiche in the back of this volume. It is hoped that these contributions will be useful to students of this important area.

Introduction

During the decade of the International Geodynamics project a large amount of work was done in the Caribbean region which revealed a number of new facts as to its constitution. However, the general tectonic history of the region and its relation to global tectonics still remains a matter of speculation.

The very term "Caribbean plate", which is commonly applied to the area may be a misnomer as it could well be described as a number of smaller "microplates". Nevertheless, it is a convenience to use the term as referring to the area bounded on the north by the Puerto Rico trench-Cayman trench-Montagua fault system, on the south by the Southern Caribbean fault system, on the east by the Lesser Antilles subduction zone and on the west by the Middle America trench subduction zone. The eastern and western boundaries are fairly well defined, the northern and southern less so. Case and Homcombe (1975) have complied a geologic-tectonic map of the Caribbean with a very complete exposition of available data.

In this report only the Caribbean sea area and the Antillean islands will be discussed. Since most of the area is under the sea, it is natural that much of the work done during the decade has been of a marine geological and geophysical nature, though a great deal of work has also been done on some of the islands.

Sub-bottom profiling together with holes drilled as part of the Deep Sea Drilling Project have considerably enhanced our knowledge of the submarine geology. In addition providing sections which greatly improve the Late Cretaceous to Neogene stratigraphy, perhaps the most outstanding discovery resulting from the drilling program has been the revelation of the nature of the sub-bottom reflectors layer "A" and layer "B". The former is now known to consist of cherts and silicified limestones of Eocene age. Layer "B" consists of basaltic dikes and flows of Late Cretaceous age, which must represent an igneous event of considerable magnitude since layer "B" is detected over all of the Venezuelan basin and probably extends over the whole Caribbean. Uplifted, on shore outcrops of layer "B" are believed to be exposed in S.W. Haiti and in Panama (Case, 1975), of layer "A" in Puerto Rico (Mattson et al, 1972) and Haiti.

Layered rocks below layer "B" consist of two units, an upper group paralleling layer "B" and a lower group at a greater inclination (Ladd and Watkins, 1977).

Recent multichannel profiling in the Venezuela basin (Biju-Duval et al., 1977) has revealed that the basin is far from homogeneous as was previously thought, different parts showing different types of sedimentary infilling and different types of basement.

The radially arranged sequence of ridges separating the basins from the Cayman Ridge to the Aves Ridge still present problems.

The Cayman Ridge appears to be an extension of eastern Cuba and dredge hauls have yielded a variety of rock types similar to those formed there. The Beata Ridge has been shown to be a strongly block faulted mass which may be an uplifted segment of Caribbean ocean floor. The fault pattern seems to be continued on to Hispaniola but the relationship between the roughly northward trending ridge of the ENE structures of Hispaniola against which it abuts remains to be explained.

The Aves Ridge is now fairly conclusively seen to be an older volcanic arc. How it relates to the present active arc and to the older Cretaceous-Eocene arcs is still obscure.

The Cayman Trough is now seen to be an east-west spreading centre from bathymetric studies (Holcombe et al., 1973), magnetic studies and exploration by the submersible "Alvin".

Compressional structures are now seen both on the north and south margins of the Venezuela Basin, in the region of the Curacao Ridge to the south and the Muertos Trough to the north. (Silver et al., 1975; Ladd and Watkins, 1977). Thus it now appears that N-S compression is an important factor in more recent tectonics. The Muertos Trench is not at present active seismically. Silver et al. report a clockwise rotation of the Bonaire block. A counterclockwise rotation of the E. Hispaniola and Puerto Rico blocks has been postulated (Weaver et al.,). Both of these would be consistent with an eastward relative motion of the Caribbean with respect to its northern and southern boundaries.

The establishment of a seismic network by the U.S. Geological Survey in Puerto Rico (in cooperation with the P.R. Water Resources Authority) and one in the Virgin Islands by Lamont-Doherty Geological Observatory has indicated a Benioff Zone dipping 45-50° from the Puerto Rico Trench southward under Puerto Rico (Carver and Tarr, 1976). It is suggested that the present motion along the Trench is left lateral transcurrent and represents a rather curious transition from the subduction zone of the Northern Lesser Antilles to the mainly transform motion along the northern plate boundary. (Schell and Tarr, 1977).

It will be observed that although considerable new data has been acquired during the Geodynamics Project. There still remain many fundamental problems unresolved. Case (1977) has enumerated some of these to which may be added the basic question of the nature and origin of the Caribbean crust.

Magnetism and Paleomagnetism

Marine Magnetic Studies

Marine magnetic anomaly patterns have been analysed for the Colombia basin (Christofferson, 1973), the Venezuelan basin (Donnelly, 1973), and the Cayman (MacDonald and Holcombe, 1978). The anomalies are moderately well defined in the central area of the Cayman trough, weakly defined in the Colombian basin and poorly defined in the Venezuelan basin. East-west trends are evident in the Colombian basin, whereas northeast directions are characteristic of the adjacent Venezuelan basin. The Cayman anomalies seem to represent the only unequivocal sea-floor spreading anomalies, with north-south trends indicative of symmetric east-west spreading in that trough as far back as about

6 m.y. B.P. Further studies are justified for the older regions of the Cayman trough, which is believed to date from the Eocene (Holcombe et al., 1973). No data on anomalies or anomaly trends are available for the Yucatan basin.

From a single seamount anomaly in the Venezuelan bain, Raff (1973) has provided the sole known estimate of a Cretaceous Caribbean paleomagnetic pole from a tectonically undisturbed terrain. That pole is close to the Cretaceous pole for North America. This result is regarded as anomalous in view of the spreading and associated relative plate motions that have taken place across the Cayman trough (MacDonald, 1974).

Paleomagnetic inclinations for cores of Cretaceous marine sediments and basalts (Lowrie and Opdyke, 1973) from the Caribbean are low relative to those predicted from the South America Cretaceous pole. The Caribbean basin paleomagnetic inclinations are in agreement, however, with inclinations from circum-Caribbean land-based sites, and with inclinations predicted from the North American Cretaceous pole.

Land-based Paleomagnetic Studies

Paleomagnetic studies of Cretaceous and Tertiary rocks in the Caribbean show widespread evidence of significant tectonic rotation. Cretaceous igneous units have received the most attention. In the Greater Antilles, Jamaica has received a great deal of attention (papers by Vincenz and co-workers; Watkins and Cambray, 1970), followed by Hispaniola, then Puerto Rico (Fink and Harrison, 1972). Relatively little is available on the paleomagnetism of the Lesser Antilles, although a program has been initiated there (Briden, pers. comm.).

In northern South America, paleomagnetic studies in the Andes and Caribbean ranges by Hargraves and co-workers and by MacDonald and co-workers have revealed ubiquitous anomalous paleomagnetic declinations suggestive of tectonic rotation.

Research in Central America by Gose and Swartz (1977) and MacDonald and Curran (1977) reveals perhaps the clearest pattern of long-term tectonic rotation, from the Cretaceous into the Tertiary. Those studies have good stratigraphic and age control, important in deciphering the mysteries of tectonic rotations. The data of Guerrero and Helsley (1976) from southern Mexico suggest that the Yuctan block may not have shared in Caribbean rotations in Cretaceous and younger time.

The relevance of Caribbean paleomagnetic data to the tectonics of orogenic zones is clear. A great variety of tectonic environments is represented: renmant arc, active trench-arc, active transform fault zones, subduction at continental margins, micro-continental fragments, and so on. No single model for tectonic rotations has appeared to explain the anomalous paleomagnetic declinations; indeed, a single explanation seems improbable. Adequate structural and tectonic models are available in the published geologic literature to explain the observed anomalous directions. What is needed is a better knowledge of local structures to demonstrate the applicability of those models. This points out a longstanding Caribbean problem: the relative scarcity of detailed geologic mapping.

The Geodynamics era in Caribbean paleomagnetic studies has provided some important basic data. The tectonic rotation problem is clearly present; its explanation presents the challenge to the next generation of studies. The marine magnetic anomalies are less well known; future research should be directed toward demonstrating the existence of, as well as age and trend of, sea-floor spreading magnetic anomalies in the major Caribbean basins.

In conclusion, with respect to Caribbean paleomagnetic achievements during the Geodynamics era, much has been accomplished, but the explanation of Caribbean evolution in terms of plate-tectonic theory is still very poorly known.

Metamorphic and Plutonic Rocks

A knowledge of the distribution, the types and ages of metamorphic rocks and the tectonic belts in which they occur is basic to an understanding of the evolution of the Caribbean arc system. Research recently done in Jamaica, Cuba, and Venezuela and now in Hispaniola has given a better idea of the rock types and the definition of paleoplate boundaries.

Draper et al., (1975) have shown that two metamorphic rock types can be recognized in the zone in the foothills of the Blue Mountain block in Jamaica. The Mt. Hibernia schist complex consists of blue schist to lower green schist assemblages; the Westphalia Schist complex consists of upper green schist to amphibolite assemblages. Mapping of the Blue Mt. Block has now been largely completed and compiled by Krijnen and Lee Chin (1978). This work has shown that indeed the metamorphic rocks are confined to the narrow belt to the south. Draper (1977) has found that relict structures within the metamorphic belt suggest that the original rocks were similar to the unmetamorphosed rocks in the rest of the Blue Mountain Inlier (i.e., the metamorphic event is post Mastrichtian and pre Eocene). Horsefield (1974) and Krijnen and Lee Chin (1978) have suggested plate tectonic models for the area which involve the northern Caribbean American plate boundary as a subduction zone in the Cretaceous.

Maresch has continued his studies of eclogite and related rocks of Late Cretaceous age in Margarita. Maresch and Abraham (1977) studied late stage variations of metamorphic conditions as a function of time in amphibolite paragonite eclogite from the north coast of Margarita. Maresch (1977) has reported that the reaction sequence parallels the geographical variation of mineral assemblages and has found that for a H_2O below some critical value omphacite bearing assemblages may be produced within matamorphic conditions appropriate to the high-P sector of the classical epidote-amphibolite facies by progressive dehydration of metabasaltic rocks.

In Cuba, Millian and Somin have published sever-

al papers giving significant new data on the important metamorphic terrains of the Sierra de los Organos, the Isla de Pinos, the Escambray mountains and the Pinas del Rio province. (Millan, 1974, Millan and Somin 1975, 1976, 1977; Somin and Millan 1974, 1977).

Terrigenous carbonates of apparent Jurassic age consitutes an important part of these areas (Millan and Somin, 1977). Somin and Millan (1977) have reported a number of K/Ar dates determined for garnet-amphibolite, meta-eclogite glaucophane schist, and amphibolite. These data are significant in determining metamorphic events and correlations with other areas.

In Hispaniola Nagle (1974) has drawn attention to paired blue schist-greenschist belts extending from Puerta Plata to the Samana peninsula. Similar rocks to those of the metamorphic terrain in Samana have been recovered from the west wall of the Mona canyon.

Considerable detailed information has been gathered by the Ministerio de Energia y Minas, Venezuela over the past 8 years on igneous and metamorphic rocks in the Cordillera de la Costa and the Perija Range in Venezuela. Much of this work relates to the nature and distribution of ultramafic rocks and their tectonic significance has been summarized by Martin (1978). She recognizes three major structural metamorphic belts of Jurassic-Cretaceous age which include eclogitic rocks together with glaucophane schists (retrograded) associated with serpentinites in fault zones, melange zones and allocthonous zones. These lithologic-structural belts were severely modified during tectonism in the Upper Cretaceous-Lower Tertiary.

An important new publication is the Mapa Tectonico del Norte de America del Sur (Martin, 1978).

Muessig (1978) has reported on the Central Falcon igneous suite of northwestern Venezuela. The rocks occur as shallow intrusives and extrusives of alkaline composition within Oligocene sedimentary rocks (Diaz de Gamero, 1978). The igneous activity occurred from 28-23 m.y. as inferred from stratigraphic constraints.

Besides the studies on metamorphic rocks reported elsewhere other petrologic studies recently made in Cuba include those of Soto (1978) and Shelugurov and Gajardo (1978). Soto (1978) has made a petrographic and chemical study of the ultramafic intrusive complexes in Cuba and concludes that they belong to essentially one complex. Shelugurov and Gajardo (1978) have reported on a trace element study of rocks from the Sierra Maestra (Oriente province).

Granitoid rocks of Late Mesozoic and early Tertiary age are well exposed in the Greater Antilles and in the Dutch and Venezuelan Antilles. The development of the Caribbean arc system has apparently not involved a continental crustal component. Thus the uncertainty of petrologic studies of intrusive rocks is reduced because the generation of magmas is less likely to be polygenetic as is the problem in considering magma genesis along continental margins.

Summaries of the progress of research on the granitoid and related basic rocks of the Greater Antilles have been given by Kesler et al., (1975), Lewis (1977), and Lewis and Feigenson (1978). Over 320 whole-rock major element analyses are now available for these rocks. Trace element data have been completed for Rb, Sr, Cu, Zn, Pb, Co, Ni and V (Kesler et al., 1977; Feigenson, 1978) and for V and Th for some samples (Donnelly and Rogers, 1978).

Initial values of the ratio Sr^{87}/Sr^{86}, now determined for a wide variety of granitoids of varying age and location, show a narrow range from 0.703-0. 705 - with a few anomalously high exceptions (Kesler, et al., 1975; Feigenson, 1978; Cheilletz et al., in press; Jones and Walker, unpublished).

From the determinations that have been made by both Rb/Sr and K/Ar isotope methods the age of crystallization and emplacement of the plutons in the Greater Antilles (excluding Cuba) ranges from at least 90 m.y. to 48 m.y. (Nagle et al., 1976 Cox et al., 1978; Kesler and Sutter, 1977; Feigenson, 1978). The intrusions along the southern Caribbean northern Venezuelan boundary were emplaced in the same time range (Santamaria and Schubert 1974; Priem et al., 1978) but intrusions younger than do not appear to be present.

Priem et al., (1978) have made a detailed Rb/Sr study of the tonalite batholith on Aruba. They find an age of 85.1 ± 0.5 m.y. for the northwestern part of the batholith and an age of 70.4 ± 2.0 m.y. for the remaining part.

Studies of the granitoid rocks are also significant to an understanding of the occurrence, associations and origin of the porphyry and vein copper deposits in the Greater Antilles (Kesler et al., 1975; Kesler, 1978). The relevance of this type of study to prospecting in Cuba has been outlined by Chejovich et al., (1977). In connection with studies of porphyry copper mineralization in northern Haiti (Cheilletz, 1976; Kachrillo, 1976; Nicolini, 1976), Cheilletz et al., (in press) have shown the short duration in time for the intrusion of the stocks and associated mineralization and alteration.

Donnelly and Rogers have continued their work on the chemical characteristics of Caribbean igneous rocks and their evolution in time. Their data include analyses for major elements, along with Rb, Sr, Ba, V, Th, Ni, Co, Zr and REE. Donnelly and Rogers (1977, 1978) recognize four major chemical groups: calc-alkaline, including high-K types; primitive island arc, and MORB, and a diabase dike-swarm group.

As concluded by Donnelly and Rogers (1978) it is unlikely that the different rock suites, in the Northeastern Caribbean at least, have been generated from mantle source regions of the same composition, but that the mantle composition has changed as the island arc system has evolved.

Our knowledge of the rock types making up the Cayman Trough and the northern portion of the Nicaraguan Rise and the evolution of these features has been increased considerably through the study of dredge hauls (Perfit, 1977, Perfit and Heezen, 1978) and through samples collected from the trough using the Woods Hole submersible Alvin.

The rocks collected from the Cayman Ridge and Nicaraguan Plateau include metamorphic rocks of green schist and amphibolite facies and granitoid rocks which bear a striking petrographic resemblance to similar rocks exposed on Jamaica.

Furthermore, K/Ar age determinations on granodiorites, and tonalites closely duplicate the ages reported for the same rock types on Jamaica (Lewis et al., 1973) indicating a closely similar Cretaceous-Palocene history for this part of the Nicaraguan Plateau and Cayman Ridge.

On the other hand the rocks dredged and those sampled by submersible from the floor of the Cayman Trench are mafic (volcanic and plutonic) and ultramafic rocks closely akin in chemical and modal composition to those recovered from the ocean floor of the Atlantic (Perfit, 1977; Stroup et al., 1978). Basalts and diabases of probable early Campanian age from the southern peninsula of Haiti are also similar in composition to the young mafic rocks from the Cayman Trench. (Sayeed, et al., 1978; Waggoner, 1978).

The interpretation is that the western portion of the North American-Caribbean plate margin was the site of subduction to the south during the Cretaceous with associated calc-alkaline igneous activity along the southern margin. A change in the stress field to left-lateral shear in the Early Tertiary crossed the Cayman Ridge to split away from the Nicaraguan Plateau and the development of a small spreading center within the rift (Perfit and Heezen, 1978).

Recent Volcanics

Petrologic and geochemical work both as part of studies of the Lesser Antilles as a whole (Brown et al., 1977, Rea and Baker, in press, Smith et. al., in press) and of specific areas: Saba (Baker et al., in press, Smith et al., unpublished), Statia (Gunn et. al., 1976, Smith et al., unpublished), St. Kitts (Baker and Holland, 1973, Baker in press), Montserrat (Rea, 1974, Rea and Baker, unpublished), Guadeloupe (Westercamp and Mervoyer 1974, Gunn et al., in press), Dominica (Wills, 1974), Martinique (Westercamp, 1975, Gunn et al., 1974, Smith and Roobol, 1974, unpublished), St. Vincent (Rowley, 1978), Carriacou (Jackson, in press) and Grenada (Arculus, 1976, 1977) have shown that in general terms the volcanic rocks north of St. Lucia are composed of typical calc-alkaline island arc suite. In contrast the more southerly islands (St. Vincent, the Grenadines, Grenada) are composed of a greater amount of basic volcanics some of which is alkaline in nature (Arculus, 1976, Brown et al., 1977). This alkaline substitute is most strongly developed on the island of Grenada which also shows the greatest geochemical differences in its calc-alkaline rocks from the rest of the arc. The Grenadines and St. Vincent appear to occupy a position intermediate between Grenada and the more northern islands (Brown et. al., 1977, Smith et. al., in press).

Sr isotope studies (Hedge and Lewis, 1971, Pushkar et al., 1973, Arulus, 1976, Stipp and Nagle, 1978, Hawkesworth, unpublished data) all seem to suggest a mantle origin for the primary magmas. The higher individual values encountered on St. Lucia, Martinique and Dominica are thought to be the result of local contamination. The predominantly basaltic southern islands show significantly higher average values (0.7048) relative to the predominantly andesitic northern islands (0.7038) with basalts of alkaline affinities having higher values than those of calc-alkaline affinities (Stipp and Nagle, 1978). However in some cases intra-island variation appears to be as great as that between different islands (Smith et al., in press). According to Hawkesworth (personal communication) the results of Sr and Nd isotopic studies on the alkaline lavas of Grenada indicate a complex evolutionary history for their parent magmas.

A characteristic feature of the Lesser Antilles arc is the occurrence of coarse grained plutonic xenoliths many of which show cumulate textures. Studies of these xenoliths by Lewis (1973), Wills (1974) and Powell (1978) have shown that they can be subdivided into two groups: ultrabasic cumulates and basic to intermediate varieties, the former are found in the basic volcanics while the latter occur within the andesitic horizons.

Comparison of the mineralogy of the ultrabasic cumulates with experimental work, and thermodynamic calculations based on mineral compositions show that the cumulates from the andesitic volcano of Statia equilibrated under pressures of around 10 kb, those from the basic volcanoes of Mt. Misery, St. Kitts, Foundland, Dominica and Soufriere-St. Vincent equilibrated under pressures of between 6 and 7 kb, while those from Grenada equilibrated at much lower pressures (2 kb) (Powell, 1978). Continued work by Powell and experimental studies by Graham (Univ. of Edinburgh) on the Grenada lavas and cumulates would help to clarify this apparent relationship between cumulate mineralogy and the composition of their host volcanics.

Volcanological studies in the region have included the studies of the 1971 eruption of Soufrière, St. Vincent (Aspinall et. al., 1973, Tomblin et al., 1971, Sigurdsson, 1977). A comparative study between the eruptions this century of Soufrière-St. Vincent and Mt. Pelée, Martinique, was published by Roobol and Smith (1975) who reached the conclusion that the 1971 magma of Soufrière-St. Vincent represented the quiet expulsion of 1902 magma which had remained high in the volcanic edifice and not an influx of new magma as proposed by Aspinall et al. (1973). Volcano-stratigraphic studies by Roobol, Smith and Rowley on a number of the active centers (Rowley, 1978, Roobol and Smith 1970, Smith and Roobol, 1978, Smith et al., 1979) have shown that although most volcanoes have erupted similar products, the proportions of these

products can vary quite drastically: e.g., the island of Saba is dominated by the extrusion of Pelean-type domes and the eruption of associated Pelean-type nueés ardentes; in contrast, Mt. Pelée, Martinique, has undergone a complex eruptive history producing a large variety of eruptive products. Perhaps the most important finding produced by these studies is the fact that Mt. Misery has also been shown to have erupted pyroclastic flows thus changing considerably the hazard potential for this volcano.

Until recently little was known of the submarine equivalents of the products of Lesser Antillean volcanoes, however the work of Sigurdsson and coworkers has done much to rectify this gap. Published work by this group includes the study of Kick 'em Jenny, an active submarine volcano in the Grenadines (Sigurdsson and Shepard, 1974) and a series of papers based on an extensive program of coring and dredging along the eastern and western margins of the Lesser Antilles (Carey and Sigurdsson, 1978, Carey et al., 1978, Sparks et al., 1978, Sparks and Carey, 1978). The results of these studies have shown that both historic and prehistoric eruptions from St. Vincent, Martinique and Dominica have produced airfall ash deposits in the Tobago Trough, pyroclastic flow deposits on the flanks of the volcano and ash turbidites in the Grenada Trough. According to Sparks and Carey (1978) in some instances subaerial pyroclastic flows have passed directly into subaqueous turbidites without passing through any intervening stages.

As can be seen from the above discussion the Lesser Antilles have been and are being subjected to a considerable amount of research on their petrology, geochemistry and volcanology. On the basis of the results obtained to date a number of ideas, some of which are in conflict, have been proposed for the origin and evolution of the Lesser Antilles arc (Brown et al., 1977, Rea and Baker, in press, Donnelly and Rogers, in press, Smith et al., in press) however many questions still need to be answered before any really definitive statements can be made.

Metallogenesis

Kesler (1978) has recently published a review of Caribbean metallogenesis and this section is a summary of his conclusions.

Whereas the metallogenic evolution of the Greater Antilles and of the Central America arcs is generally similar, consisting of manganese and massive sulphide mineralization in pre-arc time, porphyry copper deposits in the late stages of volcanism and development of laterites in areas of late Cenozoic uplift, nevertheless there are significant differences both in timing and abundance of mineral deposit types.

Massive sulphide deposits are abundant in the Greater Antilles and almost absent in Central America. On the other hand precious metal-vein and limestone-replacement mineralization is abundant in Central America and almost non-existent in the Greater Antilles. Late stage massive sulphide and manganese mineralization in the Greater Antilles differ from Central America where late stage silicic volcanic rocks contain vein ores.

These differences appear to be related to the tectonic evolution of the two arcs. In the Greater Antilles the important event was the early Cenozoic transition from north/south subduction to east-west at which time graben like valleys were developed from Puerto Rico to Cuba. In Puerto Rico, valleys seem to have localized the late stage porphyry deposits. In Jamaica and Cuba similar valleys are the site of manganese and massive sulphide mineralization.

In Central America the presence of a craton in the northern, nuclear part was the important factor in metallogenic evolution. Forming a relatively stable, elevated platform for Cenozoic volcanism it limited submarine volcanism and inhibited development of manganese and massive sulphide mineralization. It also provided a host in which pre Cenozoic intrusives could develop vein and porphyry copper mineralization and Paleozoic and Cretaceous limestones in which limestone-replacement deposits could develop.

The craton also had an effect on the composition of ore deposits. Cenozoic vein deposits on the craton are silver-rich compared to the silver-poor vein deposits of southern Central America. Lead mineralization is abundant in nuclear Central America, scarce in the southern part. Molybdenum, tungsten, tin, antimony and mercury mineralization of both vein and limestone-replacement type are present in nuclear Central America and not in Southern Central America.

The Caribbean island arcs can be placed in order of metallogenic maturity. Northern Central America is most mature with widespread intermediate and silicic volcanic rocks and associated vein deposits. Its maturity is emphasized by the presence of abundant lead and silver and late Cenozoic antimony-tungsten-mercury mineralization. Southern Central America represents an intermediate level of maturity, lacking significant deposits of lead, silver, antimony, tungsten or mercury but having late Cenozoic terrestrial, silicic volcanic rocks with vein type mineralization. The Greater Antilles are the least mature. Volcanic-associated vein mineralization are both lacking. The Lesser Antilles and the Venezuelan Antilles are at about the same level of maturity as the Greater Antilles.

The key characteristics in the maturity scale are the presence of terrestrial, intermediate and silicic volcanics with vein deposits and antimony-tungsten-mercury mineralization.

GEODYNAMICS OF CENTRAL AMERICA

Gabriel Dengo

ICAITI, Guatemala

Abstract. Most of Central America is a part of
the Carribbean plate; the rest belongs to three
other plates: North America, Nazca, Cocos. The
Motogua fault zone is the boundary between the
Caribbean and North American plates; along it the
Caribbean plate is being displaced towards the
east. The contact between the Caribbean and
Cocos plates is an active subduction zone typical
of an oceanic plate beneath a continental one.
Other contacts are more complex.

Mesozoic stratigraphy, similar on both sides of
Motogua fault, suggested they both belonged to
one unit, but it is known that the southeastern
part is younger. The geology of this part
suggests a collison between two oceanic plates
giving rise to obduction, but crustal thicknesses
and seismic velocities in Costa Rica-Panama do
not correspond to a crust of oceanic nature.

It is still obscure how different blocks arrived
at their present configuration.

Introduction

Central America has been defined as: "the
territorial and continental platform area extending from the Isthmus of Tehuantopec in Mexico, towards the east and southeast, to the Atrato low-
lands in Colombia", that is to say, comprising the
Isthmus between North and South America. In a
more strict sense of political geography, the
name Central America includes the territorial extension, land and sea, of Guatemala (and Belize),
El Salvador, Honduras, Nicaragua, Costa Rica and
Panama (Definition approved by the VIII Assembly
of the PAIGH, Guatemala, July-August, 1965). (See
Fig. 1)

Here we are dealing with geodynamical research
activities carried out between 1970 and 1978 in
Central America according to the last definition.

Most of the area is a part of the Caribbean
lithospheric plate. However, an important extension north of the Motagua fault in Guatemala is a
part of the North American plate and to the south
almost the whole of Panama and a part of Costa
Rica possibly belong to the Nazca plate. The
Middle American trench, extending parallel to the
Pacific coast of Mexico and Central America, is
the boundary between North American and Caribbean

plates on one side and the Cocos plate on the
other. So this paper is concerned with four lithospheric plates.

Activities of the Central American Geodynamics
Committee, as with those of other scientific
groups in the area, have been a collaboration
with investigations carried on by foreign groups
rather than strictly independent work. The
institutions which were most active (now or in
the recent past) in geodynamic studies are the
following:

Geological Surveying. Instituto Geográfico
Nacional, Guatemala; collaboration with New York
State University at Binghamton, Dartmouth College
Louisiana State University, the University of
Texas at Austin, Western Michigan University, and
Bundesanstalt fur Geowissenschaften und
Rohstoffe. Direccion General de Minas e
Hidrocarburos, Honduras; collaboration with the
University of Texas at Austin, U.S. Peace Corps,
UNDP, and Weslayan University. Centro de
Estudios Geotécnicos, El Salvador; collaboration
with Bundesanstalt für Geowissenschaften und
Rohstoffe. Servicio Geológico Nacional y Catastro
e Inventario de Recursos Naturales, Nicaragua.
Direccion de Geología, Minas y Petróleo and
Escuela Centroamericana de Geologia (Universidad
de Costa Rica), Costa Rica. Direccion de Recursos
Minerales, Panama, collaboration with UNDP.

Seismology. Instituto Nacional de Sismología,
Vulcanología, Meteorologia e Hidrografia, Guate-
mala; collaboration with U.S. Geological Survey,
Institute Nacional de Electrificación, and Marine
Sciences Institute of the University of Texas.
Centro de Estudio Geotécnicos, El Salvador. Universidad Nacional Autónoma de Honduras, Honduras;
collaboration with Marine Sciences Institute of
the University of Texas. Centro de Investigaciones Sísmicas, Nicaragua; collaboration with U.S.
Geological Survey. Universidad de Costa Rica,
Dirección de Geología, Minas y Petroleo, Instituto
Costaricense de Electricidad; collaboration with
the Marine Sciences Institute of the University
of Texas.

Gravity. PAIGH; collaboration with national
geographical institutes in each country.

Vulcanology. Dartmouth College; collaboration
with Instituto Geográfico Nacional and Instituto

33

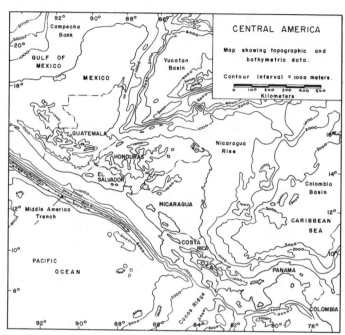

Fig. 1. Topographic and bathymetric data for the
Central American region.

Nacional de Sismología, Vulcanologia, Meteorologia
e Hidrología, Guatemala, Centro de Estudios Geo-
tecnicos, El Salvador, and Servicio Geológico
Nacional, Nicaragua. Instituto de Geofísica,
México; collaboration with PAIGH and Universidad
de Costa Rica.

Geophysics of the Middle America Trench. Marine
Sciences Institute of the University of Texas;
collaboration with Instituto Centroamericana de
Investigación y Tecnología Industrial, based in
Guatemala.

Interpretation of Satellite Images. Inter
American Development Bank; collaboration with
national geographic institutes of each country
(except Panama) and the U.S. Geological Survey.

The reasons behind this collaborative approach
are the international interest in the problems
of the area and the local scarcity of human and
financial resources.

This outside interest is strong, as reflected
by a number of conferences that have dealt with
geodynamic problems in Central America:

IV Meeting of Geologists of Central America,
Tegucigalpa, Honduras, June 23-28, 1974 (a
special session on geodynamics).

Seminar on Seismic and Volcanic Risk, San Jose,
Costa Rica, July 21-25, 1975.

Terrestrial Connections between North and
South America; Intercontinental Meeting on Science and Man, Mexico, July 3-4, 1973.

Geology, Geophysics and Resources of the Caribbean; IDOE Workshop, Kingston, Jamaica, February
17-22, 1975.

V Meeting of Geologists of Central America,
Managua, Nicaragua, February 20-26, 1977.

International Symposium on the Guatemalan
Earthquake of February 4, 1976, Guatemala City,
May 14-19, 1978.

VIII Caribbean Geological Conference, Curacao,
July, 1977.

There are a number of publications that provide
a key for the geological and tectonic knowledge
of the area, including bibliographies: Weyl
(1961) Dengo (1968, 1973), concerning metallogenesis; Levy (1979), Sykes and Ewing (1965,
Molnar and Sykes (1969), about regional seismology; Nairn and Stehli (1975), Butterlin (1977).
At the end of this paper are selected references
on the most important geodynamic topics. This
paper presents a summary without details or
discussions and detailed information concerning
investigations related to the International
Geodynamics Project may be found in the bibliographies mentioned above.

Lithospheric Plate Movements

For a number of years two different geological
provinces have been distinguished in Central
America: a northern one which includes part of
Mexico (States of Chiapas, Tabasco, Campeche,
Yucatan and Quintana Roo), Guatemala, El Salvador,
Honduras and most of Nicaragua; a southern one
which includes southern Nicaragua, Costa Rica,
Panama and western Colombia. Schuchert (1935)
called the first one Nuclear Central America and
the other the Isthmian Link. Stille (1949) called
the northern province Sapperland, in honor of Karl
Sapper, the first geological investigator of the
whole region. Lloyd (1963) and Dengo (1962)

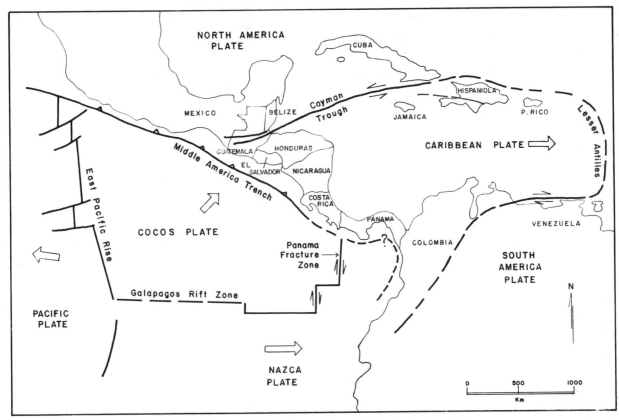

Fig. 2. Relation of Central America to the tectonic plates.

called the southern province the Southern Orogen
of Central America.

Dengo (1969) divided Nuclear Central America
into two blocks; the Mayan Block, north of the
Montogua fault, and the Chortis Block, south of
the Motagua fault. This fault trends approximately east-west across Guatemala and recent geological research has confirmed the differences between the two blocks. Currently it is accepted
that Central America consists of three lithospheric blocks with different geological his-
tories and possibly with crusts of different
composition and thickness and belonging to
different tectonic plates.

Most of Central America belongs to the Caribbean plate (Fig. 2) Tectonic models, based on
earthquake distribution, show a well defined
limit between the Caribbean and North American
plates in the Motagua fault zone, extending into
the Caribbean along the Cayman trench. However,
the boundary between these plates may be more
complex since there are other important faults
necessary essentially parallel to the Motagua
fault, such as the Polochic to the north and the
Jocotan-Chamelecon to the south (Dengo, 1968).

The Earthquake of February 4, 1976, was caused
by movements along the Motagua fault along a
length of more than 200 km. and with horizontal
sinistral displacement of more than 2 m. This
movement confirms that currently this fault is
the active border between the two plates and that
its main movement is horizontal, the Caribbean
plate being displaced to the east.

The continuation of the Motogua and related
faults towards the west and the junction of the
Caribbean and North American plates with the
Cocos plate have been studied, but remain among
the important tectonic problems yet to be solved.

It is still more difficult to fix exactly the
limits between the Caribbean, Cocos, Nazca and
South American plates. The Molnar and Sykes
(1969) model (Fig. 2) established the boundary
between the Cocos and Nazca plates along the
seismically active Panama fracture zone and the
boundary between the Nazca and Caribbean plates
along a possible faulted zone in the southern
isthmus of Panama. Studies in this latter area
reveal complex relationships not yet well understood. The north-south fracture zone off Panama
in the Pacific has been identified onshore only
in the Burica peninsula; it does not cross the
isthmus, but is cut by an ENE fault. The isthmus
of Panama is divided from the Caribbean ocean
floor to the north by an east-west trending zone
of faults. If this fault zone is the boundary
between the Nazca and Caribbean plates, then the
isthmus would belong to the Nazca plate.

The contact between the Caribbean and Cocos

plates is clearer. It is a typical collision zone of an oceanic plate with an at least partly continental one and includes an oceanic trench (Middle American Trench), a seismic zone (Benioff zone) reflecting oceanic plate subduction, and a Quaternary volcanic chain. Investigations carried on during the last decade have proved that the subduction zone is active and segmented to that the inclination and penetration velocity vary from one segment to another. The Managua earthquake of December 23, 1972 has shown different effects caused by such segmentation.

One of the most important research projects in the whole area was initiated in 1979 by the Marine Sciences Institute of the University of Texas in the Middle American trench and included bathymetry, gravimetry, seismic reflection (24 channels), and ocean bottom seismographs. During 1979 drilling beneath the sea floor was undertaken by the ship Glomar Challenger off the coast of Guatemala.

Preliminary results show a set of normal faults, dipping towards the trench near the base of the slope.

The correlation between the results of such off-shore investigations and the geology on nad is one of the most important aspects in the knowledge of the geodynamics of this area. One of the goals of the deep sea drilling was to determine if the Nicoya Complex (Nicoya Peninsula of Costa Rica) extends to the northeast parallel to the Pacific coast as inferred from magnetic studies at sea.

In some of the drill holes, basalts similar to those of Nicoya were found but correlations are difficult because, as yet, no exact ages are available. The Nicoya Complex, earlier considered to be one unit consisting mostly of oceanic basalts and radiolarities, recently has been subdivided into units ranging in age from Upper Jurassic to Eocene on the basis of the radiolarians.

Intraplate Movements

In the Central American part of the Caribbean plate, normal faults produced through extension and striking essentially north-south form, in some cases, small grabens and in other cases are reveal-ed by small volcanic cones in a line. These faults have been interpreted as resulting from the east-ward plate movement. During the Guatemalan earth-quake of February 4, 1976, the Mixco fault zone, the eastern limit of the Guatemala City graben, was reactivated as a consequence of the horizontal dis-placement of the Motagua fault.

Investigations and geological surveys during the Geodynamics Project have allowed such faults to be better traced in Guatemala, Honduras, El Salvador, and northern Nicaragua.

In several places, mainly in the Ipala graben (Guatemala), extensive faults have been accompa-nied by volcanic activity, among them small aligned cinder cones and several major stratovol-canoes of calcalkalic composition. Those close to the Caribbean coast, for example, north of Lake Yojoa in Honduras and along the Caribbean

coast of Nicaragua and Costa Rica, are of alka-line composition.

Movement Beneath Plates and Volcanic Activity

The movements beneath the Central American part of the Caribbean plate are those characteristic of oceanic plate subduction along the boundary with the Cocos plate. Such movements are reveal-ed by an active seismic zone and, indirectly, by high volcanic activity. Other types of movements beneath the plates have not been studied in this region.

Several of the volcanoes (Santiaguito, Fuego, and Pacaya in Guatemala; Izalco and San Miguel in El Salvador; San Cristobal, Cerro Negro, Momotombo and Masaya in Nicaragua; Arenal, Poas and Irazu in Costa Rica) have been studied through different methods; satellite, seismic, gravity and geochem-istry.

The parallelism of the main volcanic chain with the Middle American Trench is evidence of its tectonic origin and this volcanism is quite dif-ferent from the volcanics in the extension zones mentioned above. The alkaline volcanism along the Caribbean coast of Nicaragua and Costa Rica is a puzzle of regional tectonics since it occurs in an area where the basement is considered to be oceanic crust. Possibly this oceanic crust has grown and become partly "continentalized".

Past Movements Reflected in the Geological Record

Regional geological studies of Central America (Weyl, 1961; Dengo, 1968), prior to the Geodynamics Project, but including those of the Sixties when tectonic theory began to fluorish, accepted that the northern part of the region was one tectonic unit, both north and south of the Motagua fault. This conclusion was reached mostly because the Mesozoic stratigraphy is similar on both sides of the fault, although this is not the case for the Paleozoic stratigraphic record nor for the basement which has continental characteristics. It was known that the southern part of Central America (Costa Rica to Panama) is a younger tectonic unit, whose basement has characteristics of an oceanic origin. Nonetheless, some tectonic models of continental drift proposed, on a geometrical basis, the existence of several blocks in differ-ent positions which moved to their present posi-tions through a long history. Such models prac-tically ignored the geological record of the pro-posed blocks.

Recent investigations, especially a geological survey along the valley of the Motagua River carried out by the State University of New York at Binghamton and assisted by the Instituto Geografica Nacional de Guatemala, indicated that this area is a suture zone formed through a collision of two plates or lithospheric blocks. The block north of the Motagua fault is now part of the North American plate; the one to the south (called by some authors the Chortis or Honduras block) is a part of the Caribbean plate. Paleo-magnetic measurements suggest that the Chortis block could have been in the Pacific during the Cretaceous, but others place it originally in the Caribbean.

It has been questioned whether the Costa Rica and Panama portion has an oceanic crust since seismic investigations yield thicknesses and velocities that do not correspond to such an interpretation. Nevertheless, geological characteristics suggest that it may be an obduction zone resulting from collision between two oceanic plates between the end of the Jurassic to the Eocene.

There are many interpretations concerning the tectonic history and how the different blocks arrived at their present positions, but none of them seems to agree in a satisfactory way with the geological and geophysical information from adjacent land and sea areas.

References

Butterlin, J., Géologie Structurale de la region des Caraibes, Masson, Paris, p. 259, 1977.

Dengo, G., Estructura geólogia, historia tectonica y morfología de América Central: Centro Regional de Ayuda Tecnica, México, p. 50, 1968 (2nd ed. 1973 with bibliography until 1971).

Dengo, G., Problems of tectonic relations between Central America and the Caribbean, Gulf Coast Assoc. Geol. Soc. Trans., v.XIX, p. 311-320, 1969.

Molnar, P., and L.R. Sykes, Tectonics of the Caribbean and Middle America region from focal mechanisms and seismicity, Geol. Soc. Amer. Bull. 80, 1639-1684, 1969.

Nairn, A.E.M., and F.G. Stehli (eds.), The Gulf of Mexico and the Caribbean, V. 3, The Ocean Basins and Margins, Plenum Press, p. 706, 1975.

Levy, E., La Metalogenesis en America Central Publ. Geol. ICAITI, No. 3, p. 17-57, 1970.

Lloyd, J.J., Tectonic History of the south Central-American orogen, Am. Assoc. Petrol. Geol. Memoir 2, p. 88-100, 1963.

Sykes, L.R., and M. Ewing, The Seismicity of the Caribbean Region, J. Geophys. Res. 70, p. 5065-5074, 1965.

Selected Bibliography

Tectonic Models; Regional Aspects

Gose, W.A., and D.K. Swartz, Paleomagnetic results from Cretaceous sediments in Honduras, tectonic implications, Geology, 5, p. 505-508, 1977.

Ladd, J.W., Relative movement of South America with respect to North America and Caribbean tectonics, Geol. Soc. Amer. Bull. 87, p. 969-976, 1976.

MacDonald, W.C., The importance of Central America to the tectonic evolution of the Caribbean, Publ. Geol. ICAITI, No. 5, p. 24-30, 1976.

Perfit, M.R., and B.C. Heezen, The geology and evolution of the Cayman Trench, Geol. Soc. Amer. Bull. 39, p. 1155-1174, 1978.

Walper, J.L., and C.L. Rowett, Plate tectonics and the origins of the Caribbean Sea and the Gulf of Mexico, Gulf Coast Assoc. Geol. Socs. Trans 22, p. 105-116, 1972.

Tectonic Models; Junction of North American, Caribbean and Cocos Plates

Burkart, B., Offset of the Polichic fault of Guatemala and Chiapes, Mexico, Geology, 6, p. 328-332, 1978.

Muehlberger, W.R., and A.W. Ritchie, Caribbean-American plate boundary in Guatemala and southern Mexico as seen on Skylab IV orbital photography, Geology, 3, p. 232-235, 1975.

Tectonic Models; Junction of South America, Nazca, Caribbean, and Cocos Plates

Case, J.E., L.G. Durán, A.R. López and W.R. Moore, Tectonic investigations in western Colombia and eastern Panama, Geol. Soc, Amer. Bull. 82, p. 2685-2712, 1971.

Lonsdale, P., and K.D. Klitgard, Structure and tectonic history of the eastern Panama Basin, Geol. Soc. Amer. Bull. 89, p. 981-999, 1978.

Lowrie, A., Buried trench south of the Gulf of Panama, Geology, 6, p. 434-436, 1976.

Shagam, R., The northern termination of the Andes, in Nairn and Stehli (1975), p. 325-420, 1975.

van Andel, Tj.H., G.R. Heath, B.T. Malfait, D.F. Heinrichs and J.I. Ewing, Tectonics of the Panama Basin, equatorial Pacific, Geol. Soc. Amer. Bull. 82, p. 1498-1580, 1971.

Characteristics of the Crust

Case, J.E., Oceanic crust forms basement of eastern Panama, Geol. Soc. Amer. Bull., 85, p. 645-652, 1974.

Goosens, P.J., W.R. Rose, Jr. and D. Flores, Geochemistry of tholeiites of the Basic Igneous Complex of northwestern South America, Geol. Soc. Amer. Bull., 88, p. 1711-1720, 1977.

Schmidt-Effing, R., Alter und Genese des Nicoya-Komplexes, einer ozeanischen Paleocruste (Objera bis Eozän) im südlichen Zentralamerika, Geol. Rundschau, v. 68, p. 457-494, 1979.

Tectonics

Carr, M.J., Underthrusting and Quaternary faulting in northern Central America, Geol. Soc. Amer. Bull., 87, p. 825-829, 1976.

Dengo, G., Paleozoic and Mesozoic tectonic belts in Mexico and Central America, In Nairnand Stehli (1965), p. 283-323, 1975.

Donnelly, T.W., Geological history of the Motagua valley and of the Motagua fault system, Memoria, Simposio Internacional sobre el terremoto de Guatemala del 4 de febrero de 1976, p. 3, 1978.

Everett, J.R., and R.H. Facundini, Structural geology of El Rosario and Comayagua quadrangles, Honduras, Publ. Geol. ICIATA, No. 5, p. 31-42, 1976.

Plafker, G., Tectonic significance of surface faulting related to the 4th February Guatemala earthquake, Memoria, Simposio Internacional sobre el terremoto de Guatemala del 4 de Febrero de 1976, p. 20, 1978.

Ritchie, A.W., Jocotan fault (Guatemala), possible western extension, Publ. Geol. ICIATA, No. 5, p. 52-55, 1976.

Schwartz, D.P., L.S. Cluff, and I.W. Donnelly, Quaternary faulting along the Caribbean-North American plate boundary in Central America, Memoria, Simposio Internacional sobre el terremoto de Guatemala del 4 de febrero de 1976, p. 18, 1978.

Atlantic Continental Margin (Caribbean and Gulf of Mexico)

Arden, D.D., Jr., Geology of Jamaica and the Nicaragua Rise, in Nairn and Stehli, (1975) p. 617-661, 1975.

Case, J.E., Geophysical studies in the Carribbean Sea, in Narin and Stehli, (1975),p. 107-180, 1975.

Dillion, W.P. and J.G. Vedder, Structure and development of the continental margin of British Honduras, Geol. Soc. Amer. Bull., 84, p. 2713-2732, 1973.

Meyerhoff, A.A., and C.W. Hatten, Bahamas salient of North America, in The Geology of Continental Margins, Burk, C.A., and C.L. Drake, eds., Springer-Verlag, p. 429-446, 1974.

Perfit, M.R., and B.C. Heezen, The geology and evolution of the Cayman Trench, Geol. Soc. Amer. Bull., 89, p. 1155-1174, 1978.

Uchupi, E., Physiography of the Gulf of Mexico and the Caribbean Sea, in Nairn and Stehli (1975), p. 1-64, 1975.

Wantland, K.F., and W.C. Pusey, The southern shelf of British Honduras, guidebook, New Orleans Geol. Soc., p. 87, 1971.

Worzel, J.L., and C.A. Burk, Margins of the Gulf of Mexico, Am. Assoc. Petrol. Geol. 62, p. 2290-2307, 1978.

Pacific Continental Margin

Carr, M.J., R.E. Stoiber and C.L. Drake, The segmented nature of some continental margins, in Burk and Drake (1974), p. 105-114, 1974.

Karig, D.E., R.K. Cardwell, G.F. Moore and D.G. Moore, Late Cenozoic subduction and continental margin truncation along the Middle American Trench, Geol. Soc. Amer. Bull. 89, p. 265-276, 1978.

Ladd, J.W., A.K. Ibrahim, K.J. McMillen, G.V. Latham, R.E. von Heune, J.S. Watkins and J.C. Moore, Tectonics of the Middle America Trench offshore Guatemala, Memoria, Simposio Internacional sobre el terremoto de Guatemala del 4 de febrero de 1976, p. 16, 1978.

Shor, G.G., Jr., Continental margin of Middle America, in Burk and Drake (1974), p. 599-602, 1974.

Regional Seismology: Relations to Tectonics

Carr, M.J., and R.E. Stoiber, Geologic setting of some earthquakes in Central America, Geol. Soc. Amer. Bull., 88, p. 151-156, 1977.

Grases, J., Sismicidad e la region centroamericana asociada a la cadena volcanica del Guaternario, Tesis, Univ. Cent. Venezuela, p. 253, 1974.

Sykes, L.R., Plate tectonic framework of Middle America and the Caribbean regions, Memoria, Simposio Internacional subre el terremoto de Guatemala del 4 de Febrero de 1976, p. 11, 1976.

Seismological Studies: Guatemala Earthquake, February 4, 1976.

Espinosa, A.F., ed., The Guatemalan earthquake of February 4, 1976, a preliminary paper, U.S.Geol. Sur. Prof. Pap. 1002, 1976. This volume contains the following papers:

Dewey, James and Julian, Main event parameters from teleseismic data, p. 19-23.

Espinosa, Husid and Quesada, Intensity distribution and source parameters from field observations, p. 52-66.

Harlow, Instrumentally recorded seismic activity prior to the main event, p. 12-16.

Knudson, Strong motion recordings of the main event and February 18 aftershock, p. 24-29

Person, Spence and Dewey, Main event and principal aftershocks from teleseismic data, p. 24-29.

Spence and Person, Tectonic setting and seismicity, p. 4-11.

Fiedler, G., Das Erdbeben von Guatemala vom 4 February 1976, Geol. Rund. v. 66, p. 309-335, 1977.

Kanamori, H., and G.S. Stewart, Seismological aspects of the Guatemala earthquake of February 4, 1976, Memoria, Simposio Internacional sobre el terremoto de Guatemala del 4 febero 1976, p. 14, 1978.

Knudson, C.F., and V. Perez, Guatemalan strong motion earthquake records, Memoria, Simposio sobre el terremoto de Guatemala del 4 febero 1976, p. 24, 1978.

Langer, C.J., C.A. Bollinger and R.F. Henrisey, Aftershocks and secondary faulting associated with the 4 February 1976 Guatemalan earthquake, Memoria, Simposio Internacional sobre el terremoto de Guatemala del 4 febero, 1976, p. 26, 1978.

Matumoto, T., and G. Latham, Distribution of aftershocks following the Guatemalan earthquake of 4 February 1976 and its tectonic aspects - interplate and intraplate seismic activity, Simposio Internacional sobre el terremoto de Guatemala del 4 febero, 1976, p. 18, 1978.

Tocher, D., D. Turcotte and J. Hobgood, Seismological aspects of the Guatemalan earthquake of 4

February, 1976, Simposio Internacional sobre el terremoto de Guatemala del 4 febrero, 1976, p. 24, 1978.

Seismological Studies: Earthquake of Managua, December 23, 1972.

Brown, R.D.Jr., P.L. Ward and G. Plafker, Geological and seismological aspects of the Managua, Nicaragua, earthquakes of December 23, 1972, U. S. Geol. Sur. Prof. Pap. 838, p. 34, 1973.

Regional Petrology and Volcanology.

Horne, G.S., P. Pushkar and Shafigillah, Preliminary K-Ar data from the Laramide Series of Central Honduras, Pub. ICIATI, 5, p. 56-64, 1976.

Robin, C., and J. Tournon, Spatial relations of andesite and alkaline provinces in Mexico and Central America, Can. Jour. Earth. Sci., v. 15, p. 1633-1641, 1978.

Pilcher, H., and R. Weyl, Petrochemical aspects of Central American magmatism, Geol. Rundschau, v. 62, p. 357-396, 1973.

Weyl, R., and H. Pilcher, Magmatism and crustal evolution in Costa Rica, Pub. ICIATI, no. 5, p. 91-98, 1976.

Experimental Studies.

Dengo, C.A., Frictional characteristics of serpentine from the Motogua fault zone in Guatemala: an experimental study, Master's thesis, Center for Tectonophysics, Texas A&M University (offset reproduction), 80 p., 1978.

Metallogenesis.

Kesler, S.E., Metallogenesis of the Caribbean region, Jour. Geol. Soc. Lon., v. 135, p. 429-441, (includes Central America), 1978.

GEODYNAMIC RESEARCH IN COLOMBIA (1972-1979)

J. Emilio Ramirez[1], Hermann Duque-Caro[2], J. Rafael Goberna[3] and J. Francois Toussaint[4]

Abstract. The Island of Malpelo is part of a SW-NE trending aseismic ridge in the Pacific Ocean west of Colombia, 400 km in length, under which the Mohorovicic Discontinuity is at a depth of 16 to 19 km. To the east of the ridge is a submarine basin interrupted by the Yaquina graben and a steep coast. Geophysical investigations confirm the presence of a subduction zone beneath the Colombian coast.

In the interior, the crust attains a thickness of 45 km. In western Colombia, the geological and geotectonic structures are complex and subject to a variety of interpretations. In the Macizo de Santandar, seismic activity is concentrated in a small area at a depth of 150 km, the nature of which is under discussion. In Guajira, the crust thins to about 15 km and, as yet, its structure has received little study.

Introduction

The northwestern part of South America displays tectonic features of extremely complex nature, being the junction of three large tectonic plates, Nazca, South American and Caribbean, whose limits are not clearly defined and whose nature and structure are not well known. For this reason, Colombia, Venezuela and Panama are of special interest to both geophysicists and geologists.

With a view of studying the nature and structure of the connection between the oceanic and the continental parts of western Colombia, as well as the geology and particularly the tectonics of the contiguous regions, several different research projects were undertaken during the period of the Geodynamics Project.

The results obtained are, as yet, provisional and subject to varied interpretations. However, the conclusions which have been reached are important and point up more precisely the remaining problems which require further study and research.

---
1. Deceased.
2. National Institute of Geologic-Mineral Investigations, Bogotá, Colombia.
3. Geophysical Institute of the Colombian Andes - University of Javeriana, Bogotá, Colombia.
4. Department of Earth Sciences, National University, Medellín, Colombia.

Geophysical Investigations

In the field of geophysics the following conclusions have been reached with regard to the nasin in the Pacific Ocean between the island of Malpelo and Buenaventura, the Western Cordillera of Colombia, the region of the Macizo de Santandar in the Eastern Cordillera, the region of Antioquia-Choco and the Caribbean coast, the areas that have been the most studied.

Western Region.

a. Gravimetric studies of Case et al (1971, 1973). Case et al reached the following conclusions from their gravimetric studies of southwestern and western Colombia.

1. A continental domain was recognized, formed by the Cordillera Central, and an oceanic domain, represented by the Western Cordillera. In the latter a gravity high was described, more or less continuous from Panama to Ecuador. The authors proposed models for the evolution of this zone in terms of possible plate movements.

2. The gravimetric ridge of western Colombia was produced by an upthrown block of mafic and ultramafic crust which is located between the eastern part of the Bolivar depression and the western part of the Mesozoic eugeosyncline. This horst-like block could be related to an oceanic volcanic island ridge system (ancient Panama) which could have been emplaced by obduction.

3. The Dolores fault system, represented by the Romeral-Cauca faults, appears to spearate the oceanic crust on the west from the continental crust on the east, at least in the southern and central parts of Colombia.

4. The crustal model derived from the gravity data in the southwestern region indicates that the thickness of the crust under the Andes at about latitude 1°N is some 45 km; this thickness is somewhat greater than that deduced for farther north at 6°N. Thus the Mohorovicic discontinuity apparently dips southward along the length of the Andes to a depth of 70 km under the Peruvian-Chilean Andes (James, 1971).

b. Nariño Project. The island of Malpelo, Colombia, 500 km west of Buenaventura, forms the most prominent part of an aseismic ridge 400 km long and oriented SW-NE (Fig, 1). In the region of the island the marine sediments have thicknesses of the order of 400 to 600 m and the minimum depth

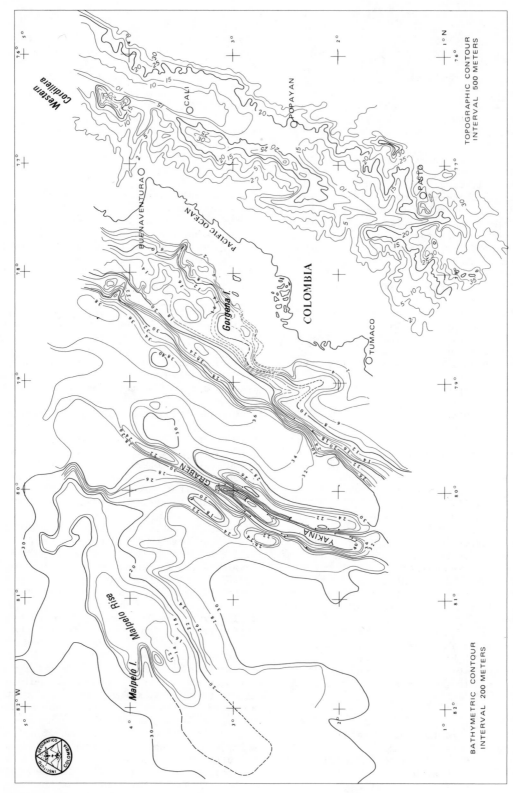

Fig. 1. Bathymetric chart of the region between Malpelo Rise and the Pacific coast of Colombia. Bathymetric and topographic contour interval is 200 meters.

to the Mohorovicic discontinuity, based on an assumed mantle velocity of 8 km/sec, varies from 16 to 19 km, in agreement with a gravity anomaly of +50 mgal found in the region.

Eastward from the Malpelo rise a submarine basin extends towards the Pacific coast of Colombia. It is characterized by an uniformly flat structure, by depths greater than 3000 m, by the well defined Yaquina graben and with an abrupt topographic junction at the coast underlain by a great thickness of sediments. This basin probably belongs to the Nazca Plate and appears to have been formed by the extension of the Cocos Ridge rather than by the rift of the East Pacific Rise. SE trending magnetic anomalies measured along 4 profiles and others previously known have not yet been correlated with the geomagnetic chronology scale. This does not mean that there has not been normal seafloor spreading, but that the age and the direction of the spreading could not be determined.

The Yaquina trench (Fig. 1) is deep, with steeply inclined walls and a level floor. It extends between the Carnegie Ridge and the continent to latitude 5°N where it loses its identity in a complex ocean bottom. High heat flow values, plus the continuous reflection profiles performed during Project Nariño I, show that the structure could be a marine version of a recent spreading ridge.

The deep Atrato-San Juan basin, which runs in a N-S direction and which can be considered to be part of the Bolivar geosyncline, appears to continue from the coastal region in front of Buenaventura towards the south, judging from the negative gravity anomalies (-90 mgal) and the thickness of sediments (up to 10 km) at the boundary between the present continental domain and the oceanic domain (Fig. 2).

The results from Nariño Projects I, II and III conducted with the support of COLCIENCIA, confirmed the subduction of the oceanic plate under the continent in the coastal plain in front of Buenaventura and Tumaco with increasing dips in an easterly direction up to more than 20° (Meyer et al, 1977; Meissner et al, 1977). Subduction along the Pacific coast of Colombia has been confirmed by refraction profiles from sea to land. by gravity measurements, by refraction profiles parallel to the structure, by heat flow and by the distribution of earthquake foci, which are near the surface east of the Yaquina graben and increase in depth to 20-30 km below the Mohorovicic discontinuity, following its dip. The thickness of the continental crust under the Andes of Nariño is calculated to be about 35 km.

The idea has arisen, confirmed by other geophysical evidence, that collision has occurred as well as subduction with possible obduction between oceanic and continental crusts. This would explain the occurrence of oceanic rocks (ophiolites and basic volcanic rocks) in nuclei of the Western Cordillera, in the Serranía de Baudó and in their extensions both in the sea and on land.

A wide zone trending in a SW-NE direction, running from the Pacific coast to the Santanders on the frontier with Venezuela, appears on the Map of Seismic Risk (Ramiréz and Estrada, 1977) as the region of highest seismic risk in Colombia. This zone crosses the three cordilleras and their most important faults. The Caribbean coast from Turbo to Santa Marta shares the aseismicity of the Caribbean Plate and the southeastern region of Colombia with that of the Guyana shield.

Massif of Santander

The massif of Santander, in view of the tectonic plates which encompass it and the swarms of earthquakes that occur, appears to be a key point in the complicated interactions of the Caribbean, Pacific and South American Plates that have played a preponderant role in the development of the northeastern corner of South America. For this reason it merited special geophysical and geological investigation. The massif of Santander combines a whole series of unique characteristics, indicating a complex history (Fig. 3).

a. General description. The massif of Santander, to the east of Bucaramanga, has its center in and occurs in a node of the Eastern Cordillera, the youngest of the Columbian cordilleras. Here the Eastern Cordillera attains its greatest elevation (5400 m, Sierra Nevada del Cocuy) and changes its direction from SSW-NNE to NNW. This massif is the most northerly of the four massifs included in this cordillera which, in order, are those of Floresta (Boyacá), of Quetame (Cundinamarca) and of Garzón (Huila). It consists of a horst, 300 km in length and 50 km in width, which during its uplift was fracured and tilted towards the west. Towards the north its two branches disappear below the valley of the river Cesar; southward the structure "founders" and has a single axis. On the eastern side it is largely bounded by the Soapaga faults and on the west by the Valley of the Magdalena with a series of N-S trending faults such as the Bucaramanga and la Salina.

In this region the Eastern Cordillera and the ridges of Perijá and of Mérida converge and diverge. Here the branching ridges of the Eastern Cordillera between Bogota and Bucaramanga end and only the eastern ridge continues between the river Chicamocha and the eastern plains. From the massif, a watershed of three river systems, rivers flow towrds the north to the Lago de Maricaibo, west to the Magdalena and the Caribbean and east to the Orinoco and the Atlantic Ocean. In this zone the river Chicamocha, which is flowing northward, makes a sharp turn and flows irregularly towards the west, eroding a canyon and abandoning its old course (Rio de Oro) as it continues to the area between Piedecuesta and Cépita.

b. Geological situation. The massif comprises a nucleus of Precambrian and Paleozoic igneous and metamorphic rocks cut by batholiths and plutons of Mesozoic age. Remnants of Devonian to Tertiary rocks cover the summits of the massif and appear in the foundered blocks. It must have been formed by a N-S trending uplift in contrast to the SW-NE orientation of the Eastern Cordillera. From the

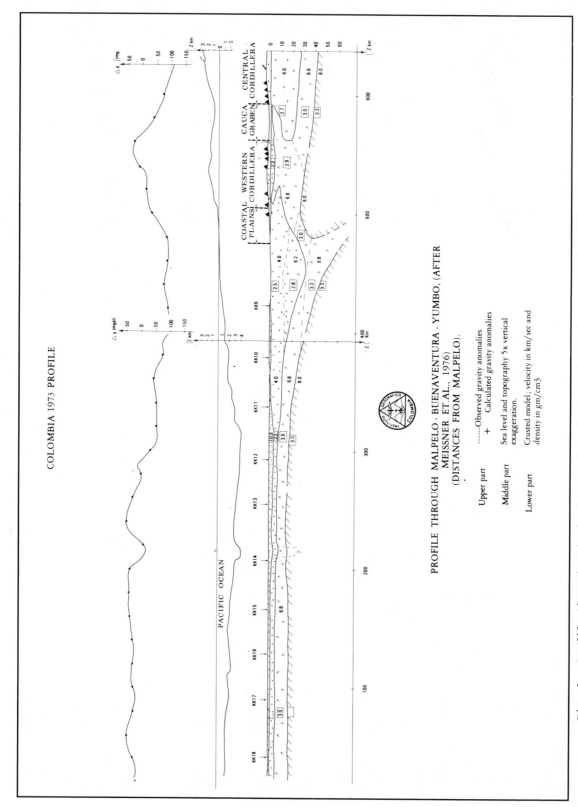

Fig. 2. Profile through Malpelo–Buenaventura–Yumbo (after Meissner et al, 1976). Distances are measured from Malpelo. ----- Observed gravity anomalies; + Calculated gravity anomalies. In the middle of the figure the topography is given at 5x vertical exaggeration. In the lower part of the figure is a crustal model: seismic velocities are in km/sec; densities are in g/cm³.

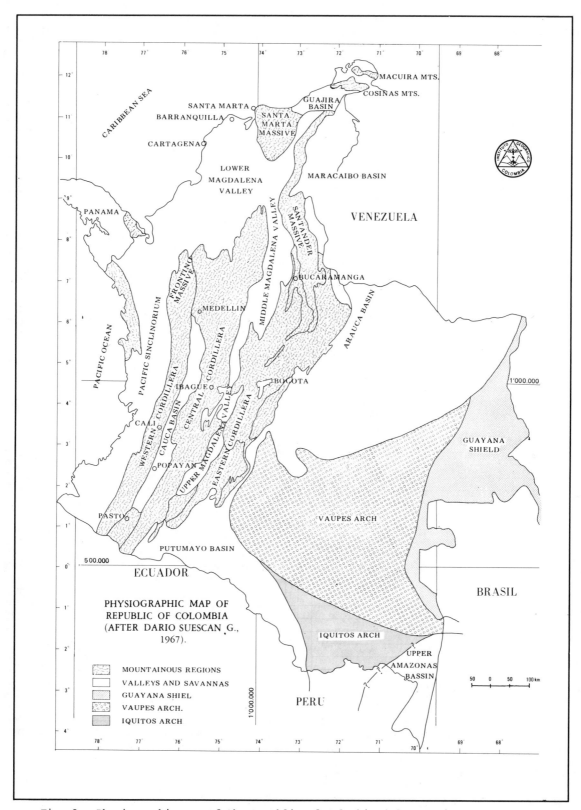

Fig. 3. Physiographic map of the Republic of Colombia (after Dario Suescan, 1967).

Paleocene onwards, vertical and horizontal movements have fractured the igneous and metamorphic rocks in a wide system of faults.

Two large faults traverse this region; the Bucaramanga and the Bocono. The Bucaramanga fault dominates the fault pattern of the massif and has been the object of more detailed studies (Ward, 1973). It is a wrench fault and runs from the extreme south of the massif towards the Caribbean coast, to the west of Santa Marta where it is called the Santa Marta fault. The presence of features which can be interpreted as "tails" or drag slices supports the idea that it is a wrench fault. Some workers (e.g. Campbell, 1965) calculate the left lateral displacement to be 110 km; others attach more importance to the vertical component of movement.

c. Seismicity. The whole of the massif is seismically active. The greatest seismicity is concentrated in a rectangle bounded by the parallels 6.5°N and 7.5°N and meridians 72.5°W and 74°W enclosing an area of 18,600 km^2. Outside this zone and in the neighboring region epicenters end abruptly along the Magdalena River. Only a few dispersed epicenters occur to the north, east, and south.

The greatest earthquake density is concentrated in a belt about 30 km wide and 130 km long which begins at 6.7°N and 72.7°W and ends at 7.2°N and 73.9°W at an angle of about 23° with the parallels in a direction ESE-WNW. At the ends of this zone two centers of higher density can be distinguished: the highest is at latitude 6.8°N and longitude 73°W, near Umpalá, and the other is some 18 km east of Bucaramanga at latitude 7.1°N and longitude 73.7°W.

The majority of the earthquakes occur at depths between 150 and 170 km and magnitudes are from 4 to 5. The deeper events are few; one was recorded at a depth of 328 km in the center of the region. Those at depths less than 70 km are more numerous, amounting to 5.6%, and occur mostly in the middle or western part of the belt. Earthquakes with magnitudes greater than 5 make up about 40% of the total and occur in the two denser centers.

In trying to correlate this seismic belt with the regional tectonics it may be observed that the belt coincides in general with the great fault of the river Chicamocha or Sogamoso. This direction defines an important tectonic zone of the Eastern Cordillera of Colombia in that the general SSW-NNE direction of the Cordillera makes a sudden sharp turn towards NNW where the seismic belt begins at latitude 6.5°N and longitude 72.5° W. Also in this region is found the highest point of the Cordillera, which reaches 5485 m elevation in the Sierra Nevada del Cocuy. Furthermore, the whole seismic belt is highly fractured and faulted in all directions. The most important faults, such as the Bucaramanga, Suarez, La Salina, Infantas, etc., have a general N-S trend, but other, lesser, faults cross the central region in various directions, especially E-W. It has not been possible to correlate the epicenters in the region with any particular fault.

Previous studies had indicated that the center of highest earthquake density has a volume of 10 km^3; recently Pennington et al (1978) reduced the volume to 4 km^3 at 6.8°N, 73,1°W and at depths between 155 and 160 km by means of portable seismographs located in the neighborhood of the massif. The average number of microearthquakes recorded in this center is 300 per month.

d. Possible interpretations. As a consequence of these phenomena it has to be admitted that deep seated tectonic forces are acting upon this region which cannot be satisfactorily explained. Among the explanations postulated by some authors are:

1. the presence of a hot point with consequent physical and chemical changes. There would be melting of rocks "at depths of 155 km, almost the end of partial fusion of andesitic rocks which in subduction ends at 160 km with temperatures of 1200°C." (Szdecsky-Kardoss, 1973).

2. the triple junction or triple point formed by the conjunction of the three plates; South American, Caribbean and Nazca.

3. sliding along the length of a subducting plate from the west and dipping towards the east, "if the point of origin of the earthquake swarms is within the subducting lithosphere and the two deep events localized in our study are on the same horizon within the plate with a N-S trend and dipping 33° eastward." (Pennington et al, 1978; Pennington, 1979).

Santa Marta Massif, Guajira Peninsula, and the Continental Margin of the Caribbean.

According to Case et al (1973) the regional gravity anomaly over the Santa Marta massif is essentially opposite to that expected in a region of topographically elevated continental crust. The Santa Marta massif, with an altitude of 5700 m and a Bouguer anomaly greater than 100 mgals, is isostatically uncompensated and under it at relatively shallow depth is an excess of mass. Similarly the positive regional anomalies in the Guajira Peninsula show that the crust is very thin and that mantle material is unusually close to the surface, perhaps at only 15 km below sea level.

In order to explain these phenomena, the high relief and the lack of isostatic balance, a mechanism can be proposed involving the overthrusting of continental crust over the adjacent crust and upper mantle of the Caribbean, either by a reverse fault or by subduction of the submarine crust. This mechanism has the advantage of explaining the right lateral displacement on the Oca fault and the left lateral displacement on the Santa Marta fault. Case (1975) described the principal structures of the Colombia Basin of the Caribbean and, on the basis of seismic reflrction and gravimetric data, proposed several mechanisms to explain the compression between the Caribbean and South America during the late Cenozoic.

Geological and Geotectonic Studies

During the period of the Porject geological field studies were intensified. The diverse interpretations of the data obtained have led to new studies and field work.

Southwestern Region. Carlos E. Acosta (1978), in his study of the southwestern region, states that fuafts and great fractures are the most important tectonic elements in the region. According to him, "the region is constituted of a series of horsts and graben trending SW-NE with some transverse horsts. The former, from west to east are: the coastal Crodillera horst, the Atrata-Pacific graben, the western Cordillera horst, the Interandean Cauca-Patía graben, the Central Cordilleran horst, the Magdalena graben, the Eastern Cordilleran horst and finally the depression of the Eastern Plains. The transverse horsts are: in the south the Nudo de los Pastos, that is known as the 'Dintel de Santa Rosa', the 'Cuchilla del Tambo' and the 'Dintel de Suarez', and farther north the 'Marmato horst'." (Acosta, 1978).

In addition to the faults and fractures, Orrego et al (1977) found proof of high pressure metamorphism on the western flank of the Central Cordillera near Jambaló (Cauca) interpreted to be the result of interplate collision during the Cretaceous. Also various Cenozoic stocks were dated radiometrically (Project 120, IGCP).

Central Zone of the Western Cordillera. According to Barrero (1977), recent stratigraphic, petrologic and geochemical studies indicate that the Western Cordillera represents a primitive island arc which formed on oceanic crust in the Lower Cretaceous. Contrary to the classic model as it is known, this island arc developed on a flexured zone in the oceanic plate, the flexure being localized on the western side of the Jurassic-Cretaceous subduction zone which bordered the continental margin of northwestern South America. The formation of a small convection cell under the zone of flexure of the oceanic plate gave rise to horizontal forces directed east and west, thus causing the migration of the subduction towards a position west of the island arc. This tectonic-magmatic process began in the Jurassic or Lower Cretaceous and ended in the Maestrichtian-Paleocene with the accretion of the island arc to the northwestern margin of the South American continent (Barrero, 1978).

Evidence for this process is the ophiolitic nature of the central region of the Western Cordillera. Field and laboratory data for the basis for the conclusion that this central region was formed from three great lithologic units which developed under different conditions, but which came together by accretion to the continental coast. These units, from oldest to youngest, are the Dagua group, the diabase group and the group of mafic and ultramafic intrusions.

The lower part of the Dagua group is formed from pelagic sediments deposited at great depths on oceanic plains far from any continental mass. In contrast, the upper part is formed from pelagic sediments containing deep and ahallow water fauna, probably deposited near a continental elevation. This great difference in conditions of deposition within a short period of time strongly suggests

that the Dagua group represents the first formation of a plate in rapid movement; this movement occurred in large part during the magnetic quiet period of the Lower Cretaceous.

The diabase group is composed of basalts and small pyroclastic and pelagic deposits. The basalts were formed during the first stage of island arc formation which, in this region, commenced in the Upper Crateceous and was located immediately to the west of a subduction zone in the arched part of the plate.

The mafic and ultramafic intrusions were formed as magmatic deposits below basaltic volcanoes, probably in the upper part of the mantle. During the Santinian to Maestrichtian this ophiolitic sequence underwent a process of folding and faulting accompanied by regional metamorphism. During the Lower Tertiary the central part of the Western Cordillera was uplifted and deeply eroded and in the Upper Tertiary was deformed and intruded during the Andean Orogeny (Barrero, 1977).

Northern Region of the Central and Western Cordillera. In the northern part of western Colombia knowledge of the principal geological features has been greatly improved (Fig. 4).
a. Recognition of magmatic events. Before 1970, the radiometric ages of only two batholiths in the Central Cordillera were known, but in the last few years radiometric dating of a large number of plutons and batholiths has been carried out, in particular through Project 120 of the International Geological Correlation Program (IGCP). With present knowledge it can be considered that magmatic activity of intermediate composition developed in the Central Cordillera during the Jurassic and Cretaceous and in the Western Cordillera during the Cenozoic, although a progressive migration from the west of the Western Cordillera towards the axis of the Central Cordillera is observed during the Tertiary. At present the great batholiths and plutons in the northern part have been dated, but only a few dates are available for the central and southern parts.

Some investigators have sought to relate the genesis of the magmatic belts to old subduction zones, but the problem of the distance from the trench to the volcanic front remains a matter of controversy (Estrada, 1971; Toussaint and Restrepo, 1976). The basic magmatism and the ophiolites have been partly described by various authors and have been dated as Cretaceous by radiometric analysis (Restrepo and Toussaint, 1975, 1978) and by paleontology (Botero et al, 1971) although many details are lacking for the differentiation of the different sequences and the volcanic environments (oceanic crust and incipient arc).
b. Recognition of metamorphic events. Until a few years ago only one metamorphic event was recognized in this region. Today at least four metamorphic events can be proved which also affect partially the region of the Rio Cauca Valley (Fig. 3). The presence of Precambrian amphibolites and gneisses has been demonstrated both on the eastern flank of the Central Cordillera (Feininger et al, 1971; Barerro and Vesga, 1976) and on the western

W

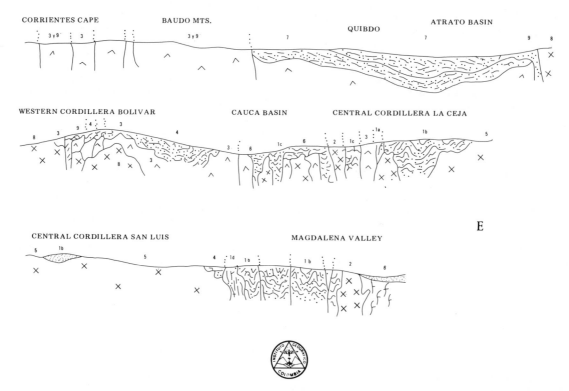

Fig. 4. Schematic geological cross section of western Colombia in the vicinity of latitude 6°N.
1. Metamorphic complex: 1a- Precambrian rocks, 1b- Paleozoic low pressure metamorphic rocks, 1c-
Cretaceous medium pressure metamorphic rocks, 1d- Paleozoic intrusive gneisses; 2. Triassic and Jurassic quartz diorite plutons; 3. Cretaceous rocks, mainly sedimentary; 4. Cretaceous marine sediments; 5. Late Cretaceous tonalite batholith; 6. Cenozoic continental sediments; 7. Cenozoic marine
sediments; 8. Cenozoic tonalite batholiths and andesite stocks; 9. Late Cretaceous basalts. (from
Restrepo and Toussaint, in preparation).

flank (Restrepo and Toussaint, 1978). As regards
the metamorphism of the Central Nucleus of the
Central Cordillera, it has been proved to be at
least Paleozoic, probably Hercynian, in age.

An important event was recorded in the region
of Arquía and Pijao, to the north of the Rio Cauca
Valley, with the discovery of medium pressure
Cretaceous metamorphic rocks (Restrepo and Toussaint, 1975). These rocks appear to continue
southward in a belt of high pressure rocks (Orrego et al, 1977). This metamorphism was related
to plate collision (subduction and/or obduction).
Finally, a late Cretaceous tp Early Tertiary metamorphic event in the central part of the Western Cordillera was dated (Barrero, 1977), but
was not recognized in the northern part where only
one important tectonic phase was observed.
c. Recognition of tectonic events. Various important tectonic phases have been recognized of which
the Hercynian, Albian, Late Cretaceous and Early
Tertiary phases are strongly marked in western
Colombia. The ages of various Cenozoic phases

(Middle Eocene and Plio-Pleistocene) were also
determined.

The Hercynian tectogenesis was related to a possible collision between North and South America
(Toussaint and Restrepo, 1976). The Albian tectogenesis was marked by the emplacement of ophiolites by thrusting (Restrepo and Toussaint, 1973)
or by subduction with formation of a melange zone
(Estrada, 1971; Barrero, 1976). According to
Toussaint et al (1979), the early Tertiary tectogenesis was principally responsible for the deformation of the Western Cordillera in the northern part and has been interpreted as the result
of a collision between the Serranía de Baudó and
the Western Cordillera. In the central zone, the
tectogenesis is probably end-Cretaceous (Barrero,
1977) causing the development of the metamorphism
of the Dagua group.
d. Tectonic evolution of the plates. There are
diverse interpretations of the evolution of western Colombia in terms of possible plate movements.
All authors suppose a migration of the old sub-

duction zones towards the west, these being successively localized in the Cauca depression, the Atrato trench and west of the Serranía del Baudó. Opinions differ as to the age of the activity in each of these. For example, Case et al (1971) and Estrada and Barrero (1974) believe that subduction was active in the Cauca area from Jurassic to the end of the Cretaceous, while Restrepo and Toussaint (1973) consider that subduction ended here at the end of the Albian.

The interpretation of the Cauca-Romeral fault system also varies with different authors: it is variously considered that the faults represent a zone of Cretaceous subduction, a zone of Albian obduction, or a system of Early Tertiary transform faults. It is probable that the Cauca area has been the scene of many tectonic phases superimposed on one another and that varied movements have taken place successively.

The basic magmatic rocks of the Western Cordillera have been interpreted as representing oceanic crust (pilcher et al, 1974) or as an incipient arc (Restrepo and Toussaint, 1975; Barrero, 1977). In the Serranía de Baudó, Goosens et al (1977) consider that the basic material that forms a large part of it represents crust lightly contaminated by an incipient arc.

The relationships between tectonism, magmatism and subduction have been studied although the conclusions at present are very speculative. It appears that for each tectogenetic phase there corresponds a jump of subduction towards the west and for each jump in subduction there is a corresponding jump in the same sense of the magmatic belt (Toussaint and Restrepo, 1978). Taking into account earlier data, these authors (1978) propose the following model to explain the evolution of this part of the Andes. During the Precambrian and Paleozoic the metamorphic basement of the Central Cordillera was formed. The discovery of Precambrian amphiobolites and gneisses on the western border of this Cordillera, documented by radiometric age determinations, indicates the presence of events earlier than Hercynian.

From Triassic to Early Cretaceous there developed a belt of tonalitic magmatism on the axis and the eastern border of the Central Cordillera which leads to the supposition of subduction activity in the Cauca region. Tensional phenomena at that time affected the Central Cordillera. During the Jurassic (?) and Early Cretaceous, oceanic material was formed (crust and seismic arc) which rode partially over the western border and the continent during the Albian. During this important tectogenetic event the medium and high pressure metaorphism developed in the Cauca zone. This tectogenetic event could have been the result of collision of an arc with the Central Cordillera which allowed a sharp jump of the subduction zone from Cauca towards the Atrato depression. This interpretation appears to be supported by the sharp jump, in the same direction, of the belt of tonalitic and quartz dioritic magmatism.

During Late Cretaceous a belt of basic magmatism developed in the Western Cordillera; dioritic in the area between the two cordillersa and tonalitic in the Central Cordillera. In addition, the Central Cordillera was affected by tensional phenomena at this time. In the northern part of the Western Cordillera the sediments of the Cañasgordas group were deposited over oceanic material from Early Cretaceous to Paleocene (?). In the central part, sediments of the Dagua group and the rocks of the Diabase group were partially intercalated.

Towards the end of the Cretaceous and the beginning of the Tertiary important tectonic events took place. In the northern part of the Central Cordillera parallel folds developed in the cover rocks and reverse faults in the oceanic basement while in the central part much stronger tectonism is indicated by metamorphism. In the Central Cordillera large transform faults developed such as the Cauca-Romeral and Palestina systems. The strong tectonism of the Late Cretaceous-Paleocene was contemporaneous with a sharp jump of the tonalitic magmatic belt towards the west which appears to indicate a sharp jump of the subduction zone from the Atrato depression towards the west of Serranía de Baudó. Perhaps this tectonic event was the result of a collision between this ridge, composed of oceanic material, and the Western Cordillera.

Cenozoic subduction caused the formation of a magmatic belt in the Western Cordillera which was partially "continentalized". Variations in the rate of subduction and in the dip of the Benioff zone could explain the progressive migration of magmatism from the eastern border of the Atrata depression towards the axis of the Central Cordillera as well as the successive tensional and compressional movements in the Cenozoic.

Caribbean Coastal Region. Duque-Caro (1977), from studies of the northwestern region of Colombia, reached the conclusion that it is composed of two great geotectonic units: the more easterly one consists of a stable platform overlying undeformed continental crust; the westerly one consists of an unstable geosyncline on oceanic crust. These two large geotectonic units are delineated by four almost parallel lineaments which mark an abrupt change in the tectonic style of each unit, the Romeral, Bolívar, Sinú and Colombia lineaments.

Taking into account the sum of their structural, sedimentary and geochronological characteristics, these four lineaments have been interpreted as ancient furrows or trenches marginal to the continent. In its early stages the Romeral lineament was a furrow and a steep scarp on the continental slope which bordered the western margin of the platform. The Bolívar lineament consists of a conspicuous reverse fault dipping eastward and with a direction parallel to the Romeral lineament, giving rise to the supposition that it represents a position of the Romeral lineament during the course of its westward migration, possibly in the Late Paleocene. Similarly the Sinú lineament, which is also interpreted in its initial phase as a trench parallel to the two former ones, is considered as the new position of the westward displacement of the Romeral lineament. Finally, the Colombia lineament, which forms the western boundary of the geosyncline with the abyssal plain, has been interpreted also as a submarine trench and as the final position of the westward migrating Romeral.

In this way progressive continental accretion towards the west along the northwestern margin of Colombia gave rise to a migration of the initial Romeral trench towards the west, occupying successively the present positions of the Bolívar, Sinú and Colombia "trenches", and a migration in the same direction of the corresponding sedimentary provinces.

Lateral compressional forces normal to the continental margin, resulting from the interaction between the oceanic crust of the Caribbean (Colombian Basin) and the South American continent, have been the principal cause of the tectonic, structural and sedimentary forms such as folding, uplifts, faulting. fractures, volcanism, plutonism, etc., of this region at least since Upper Cretaceous. The transcurrent faulting, structural flexures and geofractures form the final stage of the evolution of northwestern Colombia which took place during the Andean orogeny in the Plio-Pleistocene.

In the Baja Magdalena basin (Thery et al, 1977), in the northwest of Colombia (Fig. 3), Aquitaine Colombia drilled wells El Cabano #1 and Los Cayos #1. Radiometric measurements of cores from these wells establish two metamorphic events; one in the Paleocene and one in the Middle Cretaceous. These measurements establish contemporaneity with the Laramide orogeny and the events are related to a great uplift, in the form of an arch, which developed towards the end of the Middle Cretaceous in the NW part of the South American continent. This conclusion is in agreement with the geotectonics of this part of the world.

References

Acosta, C.E., El graben interandino Colombo-Ecua-doriano (Fosa Tectónica del Cauca-Patía y del Corredor Andino Ecuadotiano), Bol. de Geol. v. 12, No. 26, p. 63-199, 1978.

Acosta, C.E., Tectónica de fracturas en el SW de Colombia (abst.), Resumes, II Congr. Col. de Geol., Bogotá, Dic., 4-9, p. 14, 1978.

Alvarez, W., Fragmented Andean belt of northern Colombia, Geol. Soc. Amer. Mem. 130, p. 77-96, 1971.

Barrero, D., Metamorfismo regional en el Occidente Colombiano, Simposio sobre ofiolitas, Resumen, Medellín, 2p., 1974

Barrero, D., Geology of the central Western Cordillera west of Bugo and Rolandillo, Colombia, Ph.D. Thesis, Colo. School of Mines, 154 p., 1977.

Barrero, D., El origin de la Cordillera Occidental

y la migración de las zonas de subducción, Colombia (abst.), Resúmenes II Congr. Col. de Geol. Bogotá, Dic. 4-9, p. 13, 1978.

Barrero, D., and C. Vega, Mapa geológico del cuadrángulo K-9 Armero y la parte sur del J-9 La Dorada: Esc. 1:100,000, Ingeominas, Bogotá, 1976.

Botero, G., J.F. Toussaint, H. Ospina, F. Ortiz and J. Gomez, Yacimineto fosilífero de Arma, Publ. Esp. Geol. Fac. Minas, Medellín, No. 1, 13 p., 1971

Bueno, R., and C. Govea, Potential for exploration and development of hydrocarbons in Atrato Valley and Pacific coastal and shelf basins of Colombia, Amer. Asso. Petr. Geol. Mem. 25, p. 318-327, 1976.

Butterlin, J., Comparaison des caracteres structuraux des Cordilleres Sud-American Extra-Andines des Andes Centrales et des Andes Septen=trionales, Resumenes, II Congr. Latinamericano de Geología, 11-16 Nov. 1973, Caracas, p. 18-19, 1973.

Campbell, C.J., The Santa Marta wrench fault of Colombia and its regional setting, IV Carib. Geol. Conf., Trinidad, 30 p., 1965.

Case, J.E., Oceanic crust forms basement of eastern Panama, Bull. Geol. Soc. Amer. v. 85, No. 4, p. 645-652, 1974.

Case, J.E., Major basins along the continental margin of northern South America, in Continental Margins, Springer-Verlag, p. 733-741, 1975.

Case, J.E., L.G. Duran, A. Lopez and W.R. Morre, Tectonic investigations in western Colombia and eastern Panama, Bull. Geol. Soc. Amer., v. 82, No. 10, p. 2685-2711, 1971.

Case, J.E., Th. Feininger and G. Botero, Cordillera Central, Bull. Geol. Soc. Amer. v. 82, No. 10, p. 2695-2711, 1971.

Case, J.E., J. Barnes, G. Paris, H. Gonzalez and A. Viña, Trans-Andean geophysical profile, southern Colombia, Bull. Geol. Soc. Amer., v. 84, No. 9, p. 2895-2903, 1973.

Case. J.E., and W.D. MacDonald, Regional gravity anomalies and crustal structure in northern Colombia, Bull. Geol. Soc. Amer. v. 84, No. 9, p. 2905-2916, 1973.

Christofferson, E., Linear magnetic anomalies in the Colombian Basin, central Caribbean Sea, Bull. Geol. Soc. Amer., v. 84, No. 10, p. 3217-3230, 1973.

Dewey, J.W., Seismicity and tectonics of western Venezuela, Bull. Seis. Soc. Amer., v. 62, No. 6, p. 1711-1751, 1972.

Duque-Caro, H., Major structural elements and evolution of northwestern Colombia, Amer. Asso. Petr. Geol. Mem. 29, p. 329-352, 1979.

Dziewonski, A.M., and F. Gilbert, Temporal variation of seismic moment tensor and the evidence of precursive compression for two deep earthquakes, Nature, v. 247, p. 183-188, 1974.

Estrada, A., Geology and plate tectonics history of the Colombian Andes, Ph.D. Thesis, Stanford Univ., Cal., 115 p., 1972.

Feininger, Th., D. Barrero and N. Castro, Geología de parte de los departamentos de Antioquia y

Caldas, Bol. Geol. Bogotá, v. 20, No. 2, 173 p., 19t2

Gerth, H., Der geologische bau der Sudamerikamische Kordillere, Gebruder Borntraeger, Berlin, 264 p., 1955.

Gilbert, F., and A.A. Dziewonski, An application of normal mode theory to the retrieval of structural parameters and source mechanisms from seismic spectra, Phil. Trans. Roy. Soc. Lon., v. A278, p. 187-269, 1975.

Goosens, P.J., Area de islas volcánicas durante el Mesozoico Superior al Terciario Inferior a lo largo del margen continental noroccidental de la América del Sur (abst), Resúmenes, II Congr. Latinamericano de Geol., Caracas, 11-16 Nov., 1973, p. 67-68, 1973.

Goosens, P.J., W. Rose and D. Flores, Geochemistry of tholeiites of the Basic Igneous Complex of northwestern South America, Bull. Geol. Soc. Amer., v. 88, p. 1711-1720, 1977.

Hart, R.S., and H. Kanamori, Search for precursive compression for a deep earthquake, reply by Dziewonski, Nature, v. 253, p. 333-337, 1975.

Instituto Geofisico de Los Andes Colombianos, Listado de datos sismicos para la elaboracion del Mapa de Riego Sismico de Colombia, Mim. Inst. Geof., Bogota, 1977.

Irving, E.M., La evolucion estructural de los Andes mas septentrionales de Colombia, Bol. Geol. v. 19, no. 2, p. I-XIV, 1-90, 1971.

James, D.E., Andean crustal and upper mantle structure, Jour. Geophys. Res. v. 76, No. 14, p. 3246-3271, 1971.

Kelleher, J.A., Rupture zones of large South American earthquakes and some predictions, Jour. Geophys. Res. v. 77, No. 11, p. 2087-2103, 1972.

Kennett, B.L.N., and R.S. Simons, An implosive precursor to the Colombia earthquake 1970, July 31, Geophys. Jour. Roy. Astr. Soc., v. 44, p. 471-482, 1976.

Krause, D.C., Bathymetry, geomagnetism and tectonics of the Caribbean Sea north of Colombia, Geol. Soc. Amer. Mem. 130, p. 33-54, 1971.

Lonsdale, P., Ecuadorian subduction system, Bull. Amer. Asso. Petr. Geol., v. 62, No. 12, p. 2454-2477, 1978.

MacDonald, W.D., Tectonic rotation suggested by paleomagnetic results from northern Colombia, South America, Jour. Geophys. Res., v. 77, No. 29, p. 5720-5730, 1972.

Malfait, B.T., and M.G. Dinkelman, Circum-Caribbean tectonic and igneous activity and the evolution of the Caribbean plate, Bull. Geol. Soc. Amer., v. 83, No. 1, p. 251-272, 1972.

Mendiguren, J.A., Source mechanisms of a deep earthquake from analysis of worldwide observations of free oscillations, Ph.D. Thesis, Mass. Inst. Tech., 1972.

Mendiguren, J.A., Identification of free oscillation spectral peaks for 1970 July 31, Colombian deep shock using excitation criterion, Geophys. Jour. Roy. Astr. Soc, v. 33, No. 3, p. 281-322, 1973.

Mooney, W.D., R.P. Meyer, L.P. Laurence, H. Meyer and J.E. Ramirez, Seismic refraction studies of the Western Cordillera, Colombia, Bull. Seis. Soc. Amer., v. 69, No. 6, p. 1745-1761, 1979.

Mooney, W.D., R.P. Meyer, H. Meyer and J.E. Ramirez, Seismic refraction results from the Western Cordillera, Colombia (abst.), EOS, v. 59, No. 8, p. 814, 1978.

Muckelmann, R., First conclusions drawn from magnetic measurements which have been performed during a field campaign (Nariño III) from October 1977 to 1978; submitted for publication.

Orrego, A., H. Cepeda and G.I. Rodriguez, Esquitos glaucofanicos en el área de Jambalo, Cauca (Colombia), Informe inédito, Ingeominas, 14 p., 1977

Pennington, W.D., The subduction of the eastern Panama Basin and the seismotectonics of northwestern South America, Ph.D. Thesis, Univ. Wis. Madison, 1979.

Pennington, W.D., W.D. Mooney, R. Hissenhoven, H. Meyer and J.E. Ramirez, Results of a reconnaissance microearthquake survey of Bucaramanga, Colombia, Geophys. and Polar Res. Cen., Contr. No. 362, Univ. Wis. Madison, 12 p., and abst., EOS, v. 59, No. 4, p. 317, 1978.

Pichler, H., F.R. Stibane and R. Weys, Basischer magmatismus und Krustenbau im sudlichen mittelamerika Kolumbien un Ecuador, N. Jarhb. Geol. Paleon, Mh. Stuttgart, p. 102-126, 1974.

Ramirez, J.E., and L.T. Aldrich, eds., La transición océano continente en el suroeste de Colombia. The ocean continent transition in SW Colombia (Nariño-Proyecto Cooperativo Internacional, 1973), Inst. Geof., Univ. Javeriana, Bogotá, 314 p., 1977.

Ramirez, J.E., and G. Estrada, Mapa de Reisgo Sísmico de Colombia: Escala 1:1,500,000, Inst. Geof. Univ. Javeriana, Bogotá, 1977.

Restrepo, J.J. and J.F. Toussaint, Obducción cretácea en el occidente Colomiiano, Publ. Esp. Geol. No. 3, Medellín, 26 p. , 1973.

Restrepo, J.J., and J.F. Toussaint, Edades radiométricas de algunas rocas de Antioquia-Colombia, Publ. Esp. Geol. No. 6, Fac. Minas., Medellín, 24 p., 1975.

Restrepo, J.J., and J.F. Toussaint, Ocurrencia de Precambrico en las cercanías de Medellín-Cordillera Central de Colombia, Publ. Esp. Geol., No. 12, Fac. Cien., Medellín, 11 p., 1978.

Szadeczky-Kardoss, Mechanism and geochemical budget of plate tectonics, First Hungarian contribution to the work of the Int. Com. on Geod., Budapest, p. 25-33, 1973.

Thery, J.M., J. Esquevin and R. Menendez, Signification geotectonique de datations radiométriques dans des sondages de Basse Magdalena (Colombia), Bull. Cent. Rech., v. 1, No. 2, p. 475-494, 1977.

Toussaint, J.F., and J.J. Restrepo, Modelas Orogénicos de la tectónica de placas en los Andes Colombianos, Bol. Cien. de al Tierra, Medellín, No. 1, p. 1-47, 1976.

Toussaint, J.F., and J.J. Restrepo, Algunos aspectosdel desarrollo geológico del noroccidente Colombiano, II Congr. Col. de Geol. Bogotá, 1978

Toussaint, J.F., Grandes rasgos geológicos e la parte spetentriona; del occidente Colombiano, Publ. Esp. Fac. Cien. Medellín, 222 p., 1979.

van Andel, Tj H., G.R. Heath, B.T. Malfait, D.F. Heinrichs and J.I.Ewing, Tectonics of the Panama Basin, eastern equatorial Pacific. Bull. Geol. Soc. Amer. v. 82, No. 6, p. 1489-1508, 1971.

Ward, D.E., R. Goldsmith, J. Cruz and H. Restrepo, Geología de los cuadrángulos H-12 Bucaramanga y H-13 Pamplona. Departamento de Santandar, Bol. Geol. v. 21, Nos. 1-3, p. 1-132, 1973.

Weaver, J.D., ed., Geology, geophysics and resources of the Caribbean: Report of the IDOE Workshop on the Geology and Marine Geophysics of the Caribbean Region and its Resources, Kingston, Jamaica, 17-22 Feb., 1975, 150 p., 1977.

GEODYNAMICS OF ECUADOR

Carlos Eduardo Acosta

Departmento de Geosciences
Universidad Nacional de Columbia, Bogota

Abstract. The territory of Ecuador consists es-
entially of a narrow belt of the Andes forming
two large branches which are a continuation, both
north and south, of the Cordillera Occidental and
Cordillera Real (Western and Central cordilleras),
and a series of lesser heights, the Serranías
Subandinas, or Andean foothills, which form the
extension in Ecuador of the Cordillera Occidental
of Columbia and Peru.

The narrowing of the Cordillera of the Andes in
Ecuador between the Huancabamba fracture in the
south and the Guairapungo fracture in the north
is essentially due to the pressure exerted by the
Mid-Atlantic ridge in the east and the reaction of
Nazca plate subduction to the west. Compression
in this part of the Andes is greater than it is to
north or south and is displayed in the narrowing
of the Ecuadorian extension of the Magdelena Valley and the overthrusting of the Central Cordillera towards the east over the equivalents of
the Eastern Cordillera.

This phenomenon is reflected in the geodynamics
of Ecuador. Seismology identifies a typical
Benioff zone which at 450 km from the trench reaches a depth of 250 km and at about 125-150 km
depth gives rise to very active andesite volcanism.
Gravimetry discloses positive anomalies at the
coast due to the oceanic crust which prevails
there while in the Inter-Andean corridor and over
the eastern inbrication zone, strong negative anomalies are found which correspond to unusually
thick sialic crust.

Introduction

The geotectonic situation of Ecuador is not simple since the country is located, as is Colombia,
in the extreme northwest of South America and is
subjected to the action of at least four plates,
the American, Nazca, Cocos, and Caribbean, and at
the same time, to that of four ridges, Mid-Atlantic, Eastern Pacific, Cocos, and Carnegie. In the
last few years considerable work has been done on
the tectonics of the plates in this region and it
appears that gradually the problems of the geology
of this part of South America will be clarified.

Physiography

The geodynamics of Ecuador are closely related
to the position the country occupies between the
Cordillera of the Andes and the South American
continent. Situated immediately north of the
Huancabamba fracture, it does not belong to the
southern part of South America, the Brazilian
Shield, but forms the southern part of the northern Andes. If we give great importance to the
Guairapungo fracture, the Andean part of Ecuador
is also separated from the northern part of South
America, the Guyana Shield, and would constitite
an isolated fragment of the South American plate
with its own peculiar geodynamic characteristics.

To understand the constitution of the Ecuadorian
Andes, we have to compare them with the parts of
the Cordillera which bound them; to the south,
the Peruvian Andes, and to the north, those of
Colombia. The northern Andes of Peru, contained
between the Nudo de Pasco (Lat. 11°S) and the
Maranon fracture - Huancabamba deflection (Lat.
5°SO consists of three branches. The Western Cordillera runs parallel to the coast and, as its
name implies, is the branch farthest west and
nearest to the sea. It is considered to be the
divide between water flowing to the Pacific on
one side and the Amazon and Atlantic on the other.
The Central Cordillera lies between the Western
and Eastern Cordilleras. The Eastern Cordillera
borders the jungle of Amazonia and is the most
ancient and beautiful of the three and in many
places is under perennial snow. In northern Peru
this Cordillera follows a direction approximately
N30°W. Its width is about 300 km (Fig. 1).

The Colombian Andes, north of the Guairapungo
fracture, also consist of three branches, the
Western, Central and Eastern Cordilleras, which
diverge gradually towards the north. However, in
the view of many geologists (Gansser, 1950;
Radelli, 1967; etc.), there is another cordillera,
the "Coastal, or Baundó, Cordillera", which can be
followed from the Gulf of Panama to Cape Corrien-
tes where it proceeds towards the sea, a remnant
appearing as the Isla de Gorgona. Further south,
Tertiary sediments rising towards the west indi-

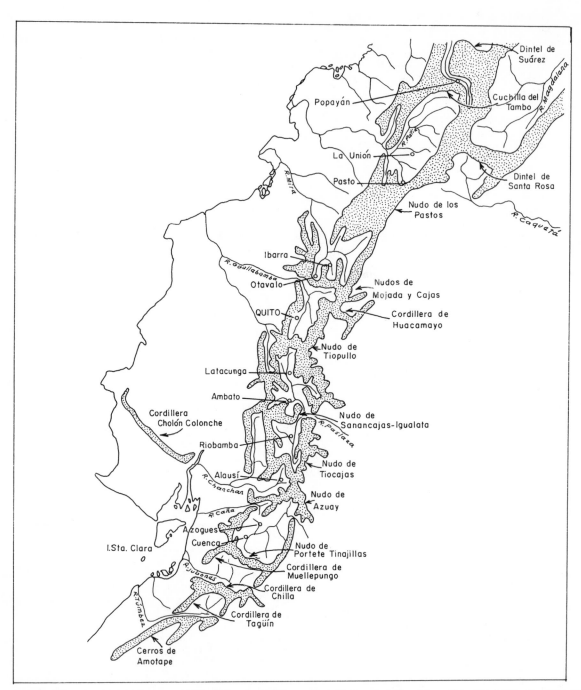

Fig. 1. Physiographic Diagram. In Ecuador, as well as in southern Colombia, the "Andes" is expressed as two cordilleras; Cordillera Occidental and Cordillera Centro-Oriental, or Real. Both of them are cut by rivers running towards either the Pacific Ocean or the Atlantic. In different places the two cordilleras are joined by transverse horsts called "nudos".

cate its continuation in the region of Tumaco whence it extends in a southwesterly direction, impinging on the western end of Ecuador in the province of Esmeraldas and probably reentering the continent in a southeasterly direction in the peninsula of Santa Elena in southwestern Ecuador. Strictly speaking the coastal uplift is not a simple mountain chain, but rather a complex alignment of various steplike uplifts (Gansser, 1950). The Coastal Cordillera in Colombia and Ecuador is composed of Mesozoic and Lower Tertiary deposits of oceanic crust type (Faldequera Group in Colombia and Pinon Group in Ecuador; Gansser, 1973).

Leaving aside the Coastal Cordillera, let us consider the three principal branches of the Andean Cordillera in Colombia. The Western , or Chocó Cordillera, trends approximately N30°E in the south of Colombia. It is the lowest of the three, but nevertheless does have some important elevations. It is quite narrow, 30-40 km. at the latitude of Patía. The Central, or Quindío, Cordillera is the one which best deserves the name of Cordillera since it is the highest and wildest of the Colombian mountain crests. In the southern region of Colombia it also is relatively narrow, 70 km. at the latitude of La Dorada. The Eastern, or Sumapaz, Cordillera, is the most important chain of the Colombian Andes from the standpoint of length, width and topography. In the southern region of Colombia its trend is almost parallel to that of the others although farther north it diverges gradually towards the east. The total width of the Andean Cordilleras at the latitude of Florencia is some 200 km., while at the latitude of Bogota it is some 340 km.

Between the northern Andes of Peru and the southern Andes of Colombia and bounded by the Huancabamba fracture zone to the south and the Guairapungo fracture to the north are located the Andes of Ecuador, an arc whose maximum curvature is found in the Guayllabamba flexure some kilometers north of Quito. North of this flexure they follow a direction similar to that of the Colombian Andes (approximately N30°E). Their length is some 200 km., extending towards the Guambuyaco - Juanambú River across the Colombia-Ecuador border. South of the Guayllabamba flexure the trend is approximately N10°E; the length some 450 km.

The mean width of the Ecuadorian Andes is about 140 km., markedly less than that of either the Peruvian or Colombian Andes, but there are parts of even lesser size. As Tschopp (1948) indicated, "at elevations no less than 500 m. above sea level, the Andes are compressed between the Brazilian shield and the Pacific to a minimum width of some 100 km. in front of the upper part of the Gulf of Guayaquil". This is in marked contrast to a maximum width of 450 km. in Colombia in the north and of some 750 km. in the south through Bolivia and Chile. Even in the region of the Magellan Straits, the mountains are approximately twice as wide as in Ecuador (Lewis et al, 1956, p. 269-270).

The Andes of Ecuador have been studied in some detail, beginning with the work of Humboldt (1816) early in the 19th Century. They consist of two principal chains, parallel and with an average separation of about 60 km. They are called the Cordillera Occidental and the Cordillera Real. The greatest separation is 85 km. in the valley of the river Paute and a minimum of 14 km. in the Zamora River valley. In places they come together in branches called "nudos" which subdivide the Interandean Passage into fourteen valleys of differing extent and relief. The geographic basins so

formed take the name of the city or principal river located in them.

The mean altitude of the Cordillera Real is some 4100 m. and that of the Occidental 4040 m. The Cordillera Real is 810 km. long, including its sinuosities and contains 10 snow fields of the first order. The Cordillera Occidental is 680 km. long and has five snow fields of the first order plus a semi-active volcano, the Guaguapichincha.

"The Eastern Colombian Cordillera appears to unite with the Central Cordillera in the vicinity of Pasto so that the depression between them becomes narrow. However, geologically, it continues south into Ecuador by way of the so-called Napo-Galeras uplift between the rivers Aguarico and Napo, by the narrow anticlinal of El Mirador, near Puyo, the Cordillera de Cutucú, and farther south, the Cordillera del Condor. A branch of the Cordillera de Cutucú goes towards the south between the rivers Santiago and Morona, is cut perpendicularly by the river Marañon in the Canon del Pongo de Manseriche, and joins in a SE direction with the Cretaceous formations of the Peruvian Eastern Cordillera" (Sauer, 1965).

The Andes of Ecuador, situated between the Colombian Andes and the Peruvian Andes, consist essentially of three Cordilleras, but with a much smaller width in Ecuador than in Colombia or Peru. The narrowing takes place not only between the Central and the Eastern, but at the expense of the size of the prolongation towards the south of the Magdelena Valley.

Stratigraphy

Many authors (Stutzer, 1934; Cizancourt, 1933; Schaufelberger, 1944; etc.) believe that the present Cordillera Oriental and Real consisted at the beginning of the Cenozoic of a single geanticlinal block resulting from the uplift of the Mesozoic geosyncline of western Colombia and Ecuador. In the Cordillera Real and particularly in the Cordillera Occidental we have Cretaceous formations of tremendous thicknesses (Bürgl, 1961, gives the Diabase Group a thickness of more than 11,000 m.). These in turn rest discordantly on older rocks the age of which has not been determined to date due to a practically total absence of fossils. It should be noted that in Colombia, as in Ecuador, it has been impossible up to now to date the metamorphic or semi-metamorphic series in the Cordillera Occidental or the Cordillera Central or its Ecuadorean extension, the Cordillera Real. It is true that in Antioquia, in the Cristalina Formation, some graptolites have been found indicating an Ordovician age. It would be highly dangerous to generalize too much and say that all these rocks are of the same age, but the uniformity of the framework of the Cordilleras, Real as well as Occidental, along the whole length of the Cauca-Patía depression and the Interandean corridor cannot be denied.

The Cordillera Real is essentially formed of what Radelli (1967) called the "Antioquian Association" in its eastern part and what previous authors have locally named the Ayurá-Montebello Group, Barragán Group, Cajamarca Group, La Ceja Geoup, Argillaceous Shale Formation, Sonson complex and in Ecuador the Metamorphic Series, the Semi-Metamorphic Series, the Paute Series, Margajitas Formation, etc.

The framework of the Cordillera Occidental in Colombia is formed essentially from the rocks of the same "Antioquian Association", the Dugua Group, in its western part, and in Ecuador by the rocks of the "Cordillera Occidental Paleozoic". These are overlain in both countries by the Mesozoic rocks of the Porphyritic Diabasic Group of Colombia and the Porphyritic and Diabasic Series of Ecuador. The ophiolitic series which are found in the Cordillera Occidental and on the inner edge of the Cordillera Central-Real deserve special mention Hubach (1957) includes them in his Faldeguera Group. In close association with them are paired metamorphic belts which include the blue schists of Jambolo in Colombia and of Machala in Ecuador.

The Interandean depression is covered with young, Tertiary and Quaternary, formations although in many places outcrops of older, more or less extensive, sedimentary, metamorphic and igneous bodies are exposed. These include the "Pre-Cambrian" rocks of Grosse (1934) which now are thought to be Jurassic, Triassic or Paleozoic, and the Porphyritic-Diabasic Group of Bürgl (1961) belonging in the Cretaceous. The same can be said for many "Andean Diorite" intrusions of Radelli (1967) in Colombia and of the granites, granodiorites and diorites of Cretaceous and Tertiary age in Ecuador.

A superficial cover of volcanic ash in the Interandean depression covers practically everything; and in the north and central part of Colombia has been named the Combia Formation with an estimated Miocene age, though it may be as young as Quaternary. In the Colombian Departamentos del Sur, the Popayán Formation (tuff layers) of Pliocene (or Pleistocene) age, contemporaneous with the Nariño Formation, is, at least in part, equivalent to the Mercaderes Formation tuffs and the tuffs of Patia. In Ecuador this ash cover has been given the Kechuan name, Cangagua, and extends throughout the Interandean Corridor.

The Cordillera Oriental is composed mainly of Mesozoic sedimentary rocks with massifs or uplifts of Paleozoic rocks within which PreCambrian rocks will probably be found. The Mesozoic rocks consist of two completely different sequences. The lower one is composed of a succession of detrital sediments which include continental redbeds, some volcanics and a few marine strata. Its thickness varies from zero to several thousands of meters and its age from Triassic to Jurassic. In contrast, the Cretaceous rocks (and Tithonian, according to Bürgl, 1961) are a marine sequence with relatively abundant faunas and no volcanics. The thickness varies and in some places attains a total of 11,000 m.

In Ecuador the Cordillera Oriental of Colombia is reduced to what has been called "Andean foothills", sub-Andean ranges", etc. A belt of uplifts, partly of complex structure, forms a prominent, but discontinuous chain which separates the High Amazonas basin from the high Andes to the west. These uplifts bring to the surface older rocks which range from lowest Pre-Macuma Paleozoic in the Pumbuiza Formation in the Cordillera de Cutucú to the Cretaceous Napo Formation along the whole length of the foothills.

The whole of the Andes, both in Ecuador and in Colombia, must be considered to be orthogeosynclinal. The Cordillera Central, with its early tendency towards emergence and its ophiolites, plays the part of a geoanticlinal uplift. The Cordillera Occidental, with its diabases, represents the eugeosyncline, whilst the Cordillera Oriental with sediments of largely continental origin takes the role of a miogeosyncline.

Structural Geology

Although one cannot deny the importance of folding in the Andes of northwestern South America, the basic structural features are associated with faulting (Fig. 2).

Cordilleras Occidental and Real (Central in Colombia. The main geological aspect of these two cordilleras is in the nature of uplifted horsts between the Pacific plains and the depressions represented by the southwest continuation of the Magdalena Valley in Colombia and the relatively low area which divides the Cordillera Real from the Sub-Andean foothills in Ecuador.

Between the horsts of the Cordillera Oriental and Real there is a large graben which in Colombia is occupied by the Cauca and Patía valleys and in Ecuador forms the Interandean Corridor. The main eastern fault is named variously according to region; Cauca, Mistrato, Cali, Patía. The eastern fault has been called the "fundamental fault of Romeral". It is accompanied along the greated part of its length by ophiolites.

Sauer (1965) notes in reference to the Interandean Corridor of Ecuador that, " the Andes rise abruptly like gigantic walls to heights of 4000 m or more above the low plains of the coast and in the east. On their ridges are superimposed volcanoes, both extinct and active, rising towards the regions of permanent snow cover. Of the former, Chimborazo, with an elevation of 6310 m, is the grandest. The separation into different partial cordilleras and longitudinal ridges, and the formation of inter-Andean depressions in the form of grabens between two horsts are characteristics which appear along the total extent of the Sierra, varying from one place to another to a greater or lesser degree."

Lewis (1976) comments, "Each Cordillera includes 5 peaks that rise to altitudes greater than 4500 m, a total of 30 such magnificent summits in about 400 km between the Colombia frontier on the north and the Cuenca Basin on the south.... the cordil-

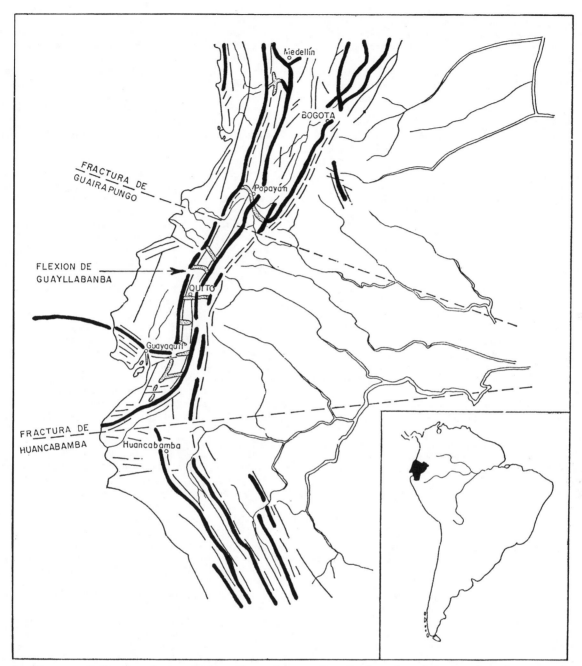

Fig. 2. Diagram of the Ecuadorian Andes. The Andean system is less broad in Ecuador, between the Huancabamba fracture in the south and the Guairapungo fracture in the north, than in Peru or Colombia. The eastern branch lies under and is compressed against the central branch and only in a few places is it possible to distinguish them as separate units. At the Guayllabamba flexure the strike of the system changes from north-south to northeast-southeast.

leras rise above the intervening depression along great fault zones. Structurally, the Intercordilleran depression may be compared to a huge graben, bounded on the east and west by fault zones that dip away from the graben at relatively high angles." Ham and Herrera (1963) add, "Between the two Cordilleras lies the Quito trench, a graben filled with Miocene to Recent volcanics and continental deposits."

Cordillera Oriental - Andean foothills. "The uplifts which produced the Andean foothills are young structures and formed at the end of the

Miocene.....Many of them are bounded on the east by large reverse faults." (Feininger, 1975).

"In Ecuador..... not all of the convergence between the oceanic and continental paltes is accomodated by thrusting on landward-dipping faults in the inner wall of the trench. A subsidiary zone of compressive folding and thrusting with landward vergence (the back-arc fold-thrust belt) is present along the boundary of the continental craton and the hot, weakened continental crust beneath the magmatic arc. In Ecuador the few kilometers of crustal shortening in this belt, most of which occurred during the Miocene, is responsible for the imbricate thrust faulting of the eastern foothills of the Andes. In Colombia, intracontinental compression uplifted the Eastern Cordillera during the late Miocene Andean Orogeny." (Lonsdale, 1978).

Volcanism

We find in Ecuador extraordinary examples of volcanoes related to different types of crustal evolution according to present day concepts of geotectonics. Ecuadorian volcanoes can be divided into two groups on the basis of their origin and composition; the marine volcanoes of the Galapagos, and the Andean volcanoes.

Volcanoes of the Galapagos. The Galapagos Islands, or Colon Archipelago, are situated between the East Pacific Rise and the west coast of Ecuador, some 100 km from each. Geotectonically they belong to the northern part of the Nazca Plate and are located about 100 km south of the Galapagos Ridge. They stand on the Galapagos Platform, which has a depth of about 1500 m and is situated near the intersection of the aseismic Cocos and Carnegie ridges. The archipelago consists of 14 principal islands and many islets, reefs and banks.

The volcanoes of the Galapagos are of the shield type with rough slopes on their upper flanks and with diameters of some 20 to 30 km, but which in some cases reach 50 km. The heights of the young volcanoes are 1500 to 1700 m above sea level which implies as altitude of 4500 m above the ocean bottom. The eruptions are predominantly of Hawaiian type with copious extrusions of lava from summits or from fissures on the upper flanks. Lavas are of pahoehoe ant aa types. The age of the volcanism and of formation of the islands has migrated from east to west. The oldest islands are of Upper Pliocen age; in these, limestones intercalated with submarne basalts have yielded fossils of Pliocene age or perhaps as old as Miocene.

Although variations have been found, in general basalts of differing types predominate. In the north-central and northeastern islands, the lavas consist of alkaline olivine basalts. In the western islands (Fernandina and Isabela) the principal lavas are tholeiitic basalts with little or no olivine, but the rocks of Ecuador volcano have close affinities with the alkaline olivine basalts.

There has been much discussion about the origin of the Galapagos Islands. The oceanic basalts suggest the probability that we are dealing with a ridge. But the alkaline nature of many of the lavas, together with the low density masses detected below the islands, cause one to consider the possibility of a mantle plume rising to the surface from a "hot spot" (Case et al, 1973; Burke et al. 1976; Hall, 1977). This would be the first known "hot spot" in the north of the Nazca Plate.

Andean Volcanoes. In continental Ecuador there are some 40 strato-volcanoes distributed unequally in the various branches of the Andean Cordillera. Of these, only 8 can be considered active, the rest being considered extinct with variable degrees of erosional destruction. The greater number are found in the Cordillera Occidental, some 15, among them Chimborazo (the highest of all the Ecuadorian volcanoes at 6227 m), Quilotoa (active), Illiniza (5266 m), Guagua-Pichincha (active), and Chiles, which marks the Colombian border. To these should be added those in Colombian territory in the Cordillera Occidental, the extension of the narrow zone of the Ecuadorian Andes, Cumbal and Azufral de Túquerres.

In the Inter-Andean Valley, there are about 10, all extinct, distributed in groups which in places cross and join the two cordilleras forming the "nudos" which bound the depressions. The location of these groups apparently is controlled by faults. In the extension of the valley into Colombia we have the only one that can be considered active, Galeras.

In the Cordillera Real there are some others, among which are Sangay (5230 m, active), Tunguragua (5016 m, active), Cotopaxi (5897 m, active, Antisana (5075 m, active) and Cayambe (5790 m, extinct). To these should be added, in the south of Colombia, Bordoncillo, near La Cocha, and farther north, the Doña Juana group; Doña Juana, Tajumbina, Animas and Petacas.

In the Sub-Andean foothills east of the Cordillera Real, and associated with a narrow zone of faulting, are the little known volcanoes, Sumaco (active), Pan de Azúcar, and Reventador (active) which form a line trending N5°W which is not parallel to the cordilleras, but which intersects the Real north of Cayambe. The strato-volcanoes end at 2°30'S and in their place is an enormous volume of pyroclastics which covers all the Andes south of Ecuador.

Petrologically, we can divide the Andean volcanics into three categories: (a) the greater part of the strato-volcanoes are composed of materials of inertmediate composition, andesites and dacites; (b) the pyroclastic material of southern Ecuador is much more siliceous than that of the north and there rhyodacites and rhyolites predominate; (c) Sumaco volcano has alkaline rocks with feldspathoids, particularly noselite and huaynite; tephrites and basanites predominate. These rocks are different from those of the other Ecuadorian volcanoes.

Although other hypotheses may be defensible, the Andean volcanism in Ecuador seems to be in accord with the model of a Benioff zone plunging from west to east. Following Lonsdale (1978), at distances between 250 and 350 km. eastward from the trench we have sources of volcanism at depths between 100 and 150 km. originating in the Benioff zone. It is logical to say that the oceanic layer basalt in passing upwards through the sialic layers of continental crust becomes contaminated, producing rocks of intermediate composition, such as andesites and dacites. Farther east the thickness of sialic crust increases and this leads us to expect that the volcanoes situated at a greater distance from the trench would have a greater content of alkaline metals and would explain the alkaline lavas of Sumaco. The enrichment in silica of the volcanics in the south perhaps may be explained by a thickening of the continental crust (giving rise to greater magmatic contamination) owing to tectonic compliations within the Huancabamba fracture zone.

Seismology

In the tectonic environment of Ecuador earthquakes of many categories are to be expected in practically the whole of the territory and neighboring areas, especially southern Colombia. These areas are now situated in a zone of very high seismicity (Dubourdieu, 1973). Unfortunately, the data available are very poor; it is highly desirable to obtain a large number of records of shallow events in order to be able to establish with certainty the existence of faults and fractures with horizontal movements which are deliniated in the topography and the bathymetry. It would be useful, above all, to have solutions of focal mechanisms which are now extremely scarce.

Earthquakes in Ecuador can be classified into five different categories:

a) Subduction earthquakes located in the Benioff zone where the oceanic lithospheric plate is dipping under the continent. Lonsdale (1978) analyzed 90 events located in a zone 180 km. in width centered on Quito and in a direction E18°S. Of these events 70, located within the Nazca Plate, are associated with a Benioff zone dipping eastward at about 23°. It is to be noted that 12 events cited by Ramírez (1975) for the same region correspond exactly to the same Benioff zone.

According to Lonsdale, "Contours of the South American Benioff zone generally parallel the trench; beneath northern Ecuador and Colombia they define a zone that dips southeast at a vertical angle of about 30°. Where this inclined zone is at a depth of 125 to 150 km. it is overlain by active andesite volcanoes. Instead of being the atypical or relict feature that van Andel et al (1971) suggested, the Benioff zone that crops out at the eastern margin of Carnegie Ridge and the Panama Basin is closer to an ideal model than that beneath Peru, where the descending slab seems to be almost horizontal (Stauder, 1975) and where there is no active volcanism."

b) Compressional earthquakes located in the concontinental crust above the Benioff zone. Of the 90 events listed by Lonsdale (1978) in the zone of Quito, 20 are continental. According to the same authors, "As at most cordilleran subduction systems, there have been many earthquakes within the continental crust overlying the Benioff zone in Colombia and Ecuador. Some of the large earthquakes have first motions appropriate for high-angle reverse faulting and indicative of east-west compressive stress across the Andes.." These events would be related to the reverse faulting which characterizes the compressive tectonics of the north west of South America (Cizancourt, 1933; Acosta, 1978b).

c) Tensional earthquakes, shallow, and located to the west of the trench in Ecuador and Colombia. With regard to these Lonsdale writes, "Although no fault-plane solutions are available and the epicenter locations are only approximate, we assume that these are tensional earthquakes at the bend in the oceanic plate as it begins to descend at the trench. This has been demonstrated for earthquakes at similar locations in the Aleutian trench (Stauder, 1968) and the Peru-Chile Trench (Stauder, 1975)."

d) Earthquakes due to "drag", usually occurring in the mantle above a subducting lithospheric plate. Although this mantle does not exist in the Andes, there is a lithospheric wedge in which seismicity can occur (James, 1978).

e) Finally we have, certainly, earthquakes due to sliding, which according to previous authors, would have great importance in Colombo-Ecuadorian tectonics. According to Lonsdale, their transcendency has been greatly exaggerated and it is not certain that sliding plays a significant role. He writes, "the evidence for the Dolores-Guayaquil megashear as a continuous transform fault is not conclusive. Few of its component faults have demonstrated strike-slip motion, and some, such as the southern Romeral fault which has lenses of Alpine-type serpentinites along it (Irving, 1975), usually have been interpreted as major dip-slip faults. The Guayaquil fault of Marchant (1961) was inferred mostly from the shape of the coastline. There is no band of shallow-focus earthquakes along this postulated transform fault, although such continental transform as California's San Andreas typically have very shallow (less than 15 km.) seismicity. The earthquakes in Colombia that Case et al (1971) and Campbell (1974a) suggested were associated with the transform faults have focal depths of 70 km. and 290 km., are clearly mantle earthquakes, and, because they occur in a southeasterly dipping Benioff zone, are more reasonably attributed to stresses within the downgoing slab subducted at the Colombia Trench."

Gravimetry

There is now a Gravity map of Ecuador, compiled by Dr. Tomás Feininger (1977) of the National Technical School at a scale of 1:1,000,000 with contour intervals of 10 milligals. On it appear data of great importance for geotectonic and geodynamic interpretation in this country.

In the east, Bouguer anomalies are negative. They vary from -100 to -150 milligals at the foot of the Andes and their adjacent foothills and regionally become less negative with increasing distance from the mountains, reaching values between 0 and -10 milligals. La Sierra and its eastern sub-Andean foothills are chiefly characterized by strongly negative Bouguer anomalies. The axis of the more intensely negative anomalies is slightly displaced to the west of the longitudinal axis of the topography of the Cordillera. Along the length of the Sierra the minimum values increase progressively towards the south from -230 milligals in Tulcan on the Colombian frontier to - 292 milligals, the greatest negative value in the country, at Latacunga. Farther south, the axis of most negative values consists of a series of lows with values of about -270 milligals in Cuenca and approximately -250 in Saraguro. The data from the more southerly region, 40 km south of Loja, give minimum values 10 milligals less negative than those at Tulcán.

Along the coast the total relief in the Bouguer gravity field is greater than in the other provinces. The variation exceeds 300 milligals, from -150 at 55 km south of Guayaquil, to +162 18 km west of Daule. This last value is the strongest positive one in Ecuador. The positive anomalies of the Ecuadorian coast form a narrow and diffuse band extending from Colombia across the northern two thirds of the coast and passing out to sea in the Gulf of Guayaquil.

Crustal Structure of Ecuador (from Feininger, 1977). "Two features in the Bouguer gravity field of Ecuador are predominant: the positive anomalies along the coast, in places more than +160 milligals, and the great belt of strong negative anomalies which coincided with the Sierra. These and other features of the gravity field permit a general, though preliminary, interpretation of crustal structure in Ecuador.

The western base of the Andes, at least north of the latitude of Guayaquil, follows the boundary between the negative anomalies of the Sierra and the positive anomalies of the coast. This boundary coincides with the eastern border of the high gravity belt of western Colombia, interpreted by Case et al (1971) and Case, Barnes et al (1973) as marking the boundary in Colombia between oceanic crust on the west and continental crust to the east. This interpretation appears to be equally valid in Ecuador and is supported both by the present Bouguer anomaly map and by geological field observations along the coast. The strong positive anomalies coincide with outcrop areas of the Piñón Formation (or the Basic Igneous Complex of Goosens and Rose, 1973) of Cretaceous age or where it is overlain by a thin cover of younger sedimentary rocks as in the quadrilateral formed by the cities and towns of Guayaquil, Babahoyo,

Balzar and Paján̄. The Piñón Formation consists of basalt (with pillows), diabase and gabbro, and peridotite. The chemical composition, lithology and structure of the formation point to an oceanic origin. The Piñón Formation of the coast is a remnant of ocean bottom of Cretaceous age. This remnant was preserved and now crops out due to its having been isolated by a jump towards the west of the subduction zone of the continental border in Lower Tertiary time: from a former trench located at the site of the present Cordillera Occidental of the Andes, to the present trench off the Ecuadorean coast (Feininger, 1975). The strictly oceanic nature of the Piñón Formation of the coast is evidenced by the great extent and magnitude of the positive Bouguer anomalies over its outcrops. For example, the positive anomalies stronger than +100 milligals cover an area of more than 4000 km^2 in the neighborhood of Daule. It is clear that continental crust cannot underlie the Piñón Formation along the Ecuador coast. On the other hand, the negative Bouguer anomalies of the coast coincide chiefly with three deep basins filled with sedimentary rocks mainly of Miocene age or younger. From north to south these are the Borbón basin to the south and east of Esmeraldas, the Manabí-Esmeraldas basin between Rosa Zárate and Portoviejo and the Progreso basin to the south and southeast of Guayaquil.

The belt of negative Bouguer anomalies which coincides in space with the Cordillera of the Andes is the reflection of the root of low density rocks of these mountains. The base of the crust below the Andes from the north of Colombia to southern Chile has been described as a canoe with progressive deepening from 30 to 40 km. in latitude 6°N to 70 km. below the central Andes at the junction of Peru, Bolivia and Chile followed by progresively lesser depths towards a point some 40 or 50 km. into Patagonia (James, 1971; Case, Barnes et al, 1973). The Andean Bouguer anomalies of Ecuador are not in accord with such a simple interpretation. Although the anomalies become progressively more negative towards the south from the Colombian frontier, they reach their minimum value in Latacunga which suggests that the deepest part of the root of the Andes lies beneath that city. South of Latacunga the negative Bouguer anomalies lose strength and in Alausí are 10 milligals weaker than in Tulcán. Between Alausí and the Peruvian border the anomalies vary from -220 to -260 milligals and apparently nowhere reach values as negative as those at Latacunga although data are lacking south of latitude 4°20'S. Since the width of the Andean Cordillera is uniform through the length of Ecuador, more or less 140 km., the longitudinal variations in the negative Bouguer anomalies chiefly reflect the depth of the Andean root and perhaps to a lesser degree, variations in the density of the rocks of the Cordillera. In summary, the Andean root becomes deeper from the Colombian border to Latacunga and thereafter becomes less deep, maintaining a more or less uniform depth from the vicinity of Riotamba to the Peruvian border in the south.

The Bouguer anomaly field in the east is characterized by strong undulations. Its weak negative values away from the Sierras are typical of continental crust with a more normal thickness of 30-35 km. The progressively more negative anomalies toward the west, especially those near the Sierra, demonstrate the influence of the deep Andean root of low density rocks. However, the smooth and uniform gradient across the whole of the east suggests that a part of it is caused by a general and progressive increase in the thickness of the continental crust from east to west. The increase in thickness may be the result of the emplacement of granite at depth from magmas derived from the underlying Benioff zone, the piling up of thrust slabs toward the east or a combination of the two. The numerous reverse faults with westerly dips in the eastern base of the Andes and the adjacent foothills shown on the geological map of Ecuador (Servicio Nacional de Geología y Minería, 1969) support the possible importance of tectonic thickening of the crust by overthrusting in the western part of eastern Ecuador.

Cretaceous and Tertiary sedimentary rocks which, with the exception of the sub-Andean foothills, are almost horizontal, underlie the eastern region. The variations in the Bouguer gravity field in the east do not derive from these slightly disturbed rocks, but rather from the enormous variety in the underlying basement rocks. These include granulitic metamorphic rocks of Precambrian age, Paleozoic sedimentary rocks, and marine and non-marine sedimentary rocks and volcanics of Triassic and Jurassic age (Feininger, 1975). As a consequence, gravity highs have been of little use in exploration form petroleum which occurs in the Cretaceous sedimentary rocks."

Geotectonics

Although Ecuador is in a strategic geographic situation and for that reason is subject to many influences, "The evidence from field geology, seismicity, and first-motion studies seems to be that in the equatorial region the Ecuador Trench is the only significant boundary between the Nazca and South American plates...." (Lonsdale, 1978) Thus the interaction between these two plates is most important in explaining the geotectonic phenomena of South America.

The geotectonics of Ecuador are entirely explained in terms of the pressure exerted by the Mid-Atlantic Ridge and the opposing reaction of the Nazca Plate. The Cordillera of the Andes is subjected in this region to a tremendous compression and its width must be considerably reduced. By means of faults with vertical components, the horizontal pressure is translated into a double phenomenon - taphrogenesis and overthrusting.

Although it has been written many times that tectonic trenches have their origin in gravity subsidence between normal faults, it can be easily understood that, as an effect of compression, a

series of blocks bounded by reverse faults can be formed into horsts and grabens. Such appears to be the general case in the Andes of Colombia and Ecuador (Cizancourt, 1933).

The inter-Andean graben would not be a case of subsidence, but simply a case of uplift retarded in comparison with the two cordilleras which bound it. The same compression phenomenon, exaggerated even more and aided by enough oblique faults, can produce overthrusts, sometimes considerably imbricated, as in the Cordillera Oriental (Colombian) in Ecuador. The Ecuadorean continuation of the Magdalena Valley has been in a compressed between blocks of the Cordillera Real pushed towards the east and the rocks of the Cordillera Oriental forced under them. This, in turn, explains the more or less complete disappearance of the Cordillera Oriental in Ecuador and the great negative gravity anomaly which is found over the eastern slope of the Cordillera Real.

In their report on the Nariño Project, Ramírez and Aldrich (1977) write of "a trench partly filled which is believed to be the analog or an extension of the Peru-Chile trench." In fact, already reference is made openly to the Ecuadorean trench and the Colombian trench (Lonsdale, 1978; Lonsdale and Klitgard, 1978). This is the beginning of a typical Benioff zone "very close to an ideal model" dipping some 23° with andesitic volcanism originating at a depth of about 150 km. and distant earthquakes at up to 650 km. depth.

The Ecuadorean Andes are bounded both north and south by fault zones which certainly represent continuations on the continent of great oceanic fracture zones (Loczy 1970b; Gansser, 1973; Meyerhoff and Meyerhoff, 1974). Lonsdale's opinion (1978) that the continuity of the ocean fracture zones into the continent is purely coincidence does not seem logical. The southern limit is quite clear and is formed by the so-called "Huancabamba deflexion" or the "Amotape transverse structure". Gansser (1973) calls it "the major transverse form in all South America which divides the continent into a northern and a southern block and which forms a structural lineament 'cut by a knife'." It is formed by the most complicated interlacing of various structural features and of events of diverse ages. According to Loczy (1970a), this structure with its east-west transverse faults appears to exhibit a close connection with the Romanche Fracture Zone in the Atlantic and the Galapagos Fracture Zone and the Carnegie Ridge in the Pacific.

The northern boundary is less spectacular although further studies should be able to clarify it. Attention is called in the first place to the alignment of the rivers Patía, Juanambú, Mocoa amd Caquetá and the apparent disappearance of the Cordillera Oriental in front of Puerto Limón. Although a structural lineament (cut by a knife) is not discussed, it is interesting to note, following Vergara y Velasco, (1901) that there is a fracture in the cordillera at this point. He writes,"The Cordillera del Quindío or Central is lowered suddenly in Guairapungo (2600 m.), the lowest pass to the south of Antioquia, open laterally over the huge fissure of Juanambú which extends to that of Minamá (Patía) and matches the breach through which the Patía reaches the sea. It is a pity that there is no seismic evidence nor bathymetric charts to confirm the existence of a zone of transform or transcurrent faults in this region, but the geomorphology and the tectonics (the positioning of the Cordillera Oriental against the Central) cause one to suspect a transverse feature of great importance.

References

Acosta Arteaga, C.E., El Graben Interandino Colombo-Ecuadoriano, Universidad Industrial de Santandar, Bucaramanga, Colombia, Boletín de Geología, v.12, no.26, p. 63-119, 1978a

Acosta Arteaga, C.E., Tectónica de Fracturas en el SW de Colombia (abst.), II Congreso Colombiano de Geología, Resúmenes p.14, Bogotá, Dec. 4-9, 1978b.

Bürgl, H., Historia Geológica de Colombia, Revista de la Aacdemia Colombiana de Ciencias Exactas, Físicas y Naturales, v.XI, no.43, p. 137-191, 1961.

Burke, K.C., and J.T.Wilson, Hot Spots on the Earth's Surface, Sci. Amer. v. 235, no.2, p. 46-57, Aug. 1976.

Campbell, C.J., Ecuadorean Andes, in Mesozoic-Cenozoic Orogenic Belts, Data for Orogenic Studies, Geol. Soc. Lon. Spec. Pap. 4, p. 725-732, 1974.

Case, J.E., L.G.Duran, A.Lopez, and W.R.Moore, Tectonic investigations in western Colombia and eastern Panama. Geol.Soc.Amer.Bull.,V.82, p. 2685-2904, 1971.

Case, J.E., J. Barnes, G.Paris, H.González and A.Viña, Trans-Andean Geophysical Profile, Southern Colombia, Geol.Soc.Amer.Bull., v.84, p.2895-2904, 1973.

Case, J.E., S.L.Ryland, T. Simkin and K.A. Howard, Gravitational Evidence for a Low-Density Mass beneath the Galapagos Islands, Science, v.181, p. 1040-1042, 1973.

Cizancourt, H. de, Tectonic Structure of Northern Andes in Colombia and Venezuela, Amer. Ass. Pet. Geol. v. 17, p. 211-228, 1933.

Childs, O.E., and B.W.Beebe, Backbone of the Americas - Tectonic History from Pole to Pole, A Symposium, edited by O.E.Childs and B.W.Beebe, Amer. Asso. Pet. Geol. Tulsa, Okla, 320 p., 1963.

Dubourdieu, G., Carte Sismique de Monde, Laboratoire de Géologie du College de France, Paris, 1973.

Faucher, B., and E.Savoyat, Esquisse Géologique des Andes de l'Equateur, Revue de Geographie Physique et de Géologie Dynamique, v. XV, Fasc. 1-2, p. 115-142, Paris, 1973.

Feininger, T., Origin of Petroleum in the Oriente of Ecuador, Amer.Asso.Petrol.Geol.Bull., v.59, p. 1166-1175, 1975.

Feininger, T., Mapa Gravimétrico de Anomalías Bouguer simples del Ecuador, Escala 1:1,000,000,

Escuela Politecnica Nacional, Quito, 1977.

Gansser, A., Geological and Petrographical Notes on Gorgona Island in Relation to Northwestern South America, Bull. Suisse de Min. et Petr., v. 30, p. 219-237, 1950.

Gansser, A., Facts and Theories on the Andes, Jour. Geol.Soc. London, v.129, p. 131, 1973.

Grosse, E., Acerca de la Geología del Sur de Colombia, Informe rendido al Ministerio de Industrias sobre un viaje por la Cuenca del Patía y el Departamento de Nariño, Comp. Est. Geol. Ofic. en Colombia, Tomo III, p. 139-235, Figs. 49-85, Dos croquis en colores, 1934.

Hall, M.L., El Volcanismo en el Ecuador, XI Asemblea General del IPGH y reuniones panamericanas de Consulta Conexas, Biblioteca Ecuador, Quito, 120 p., 1977.

Ham, C.K., and L.J.Herrera, Jr., Role of Subandean Fault Systems in Tectonics of Eastern Peru and Ecuador, in Backbone of the Americas, Childs and Beebe, eds., p. 47-61, 1963.

Hubach, E., Contribución a las Unidades Estratigráficas de Colombia, Servicio Geológico Nacional, Informe no. 1212, 1957.

Humboldt, A., von, Vues des Cordilleres et Monuments des Peuples Indigènes de l'Amérique, v.2, Librairie Greque-Latine-Allemande, Paris, 1816.

Humboldt, A., von, Cosmos, Ensayo de una descripción física del mundo, Vertido al castellano por Bernardo Giner y José de Fuentes, Tomo IV, 634 p., Imprenta de Gaspar y Roig, Editores, Madrid, 1875.

Irving, E.M., La Evolucion Estructural de los Andes más Septentrionales de Colombia, Instituto Nacional de Investigaciones Geológico-Mineras Bogotá, Bol. Geol. v.XIX, no.2, p. 1-90, 1971.

James, D.E., Subduction of the Nazca Plate beneath Central Peru, Geology, v.6, p. 174-178, 1978.

Lewis, G.E., Andean Geological Province, in Handbook of South American Geology, Geol.Soc.Amer. Memoir 65, p. 269-276, 1956.

Loczy, L. de, Tectonismo Transversal na America do Sul e suas relaçoes Genéticas com as Zonas de Fractura das Cadeias Meio-Oceânicas, An.Acad. Sci. Brasil, Ciencias 42 (2), p. 185-205, 1970a.

Loczy, L. de, Role of Transcurrent Faulting in South American Tectonic Framework, Amer.Assoc. Petrol.Bull., v.54, no.11, p. 2111-2119, 1970b.

Lonsdale, P., Ecuadorian Subduction System, Amer. Asso.Petrol.Geol.Bull, v.62, no.12, p. 2454-2477, 1978.

Lonsdale, P., and K.D.Klitgord, Structure and Tectonic History of the Eastern Panama Basin, Geol. Soc.Amer.Bull., v.89, no.7, p. 981-999, 1978.

Marchant, S., A Photogeological Analysis of the Structure of the Western Guayas Province, Ecuador, Geol.Soc.Lon.Quat.Jour, v.117, p.215-231, 1961.

Meyerhoff, A.A., and H.Meyerhoff, The New Global Tectonics: Major Inconsistencies, Amer.Assoc. Petrol.Geol.Bull., v.56, no.2, p. 269-336, 1972.

Radelli, L., Geologie des Andes Colombiennes, Travaux du Laboratoire de Geologie de la Faculte des Sciences de Grenoble Mem.no.6, 457 p., 1967.

Ramírez, J.E., Historia de los Terremotos en Colombia, Instituto Geográfico Agustin Codazzi, Segundo Edicion, 250 p., 1975.

Ramírez, J.E., and L.T.Aldrich, La Transición Oceano-Continente en el Suroeste de Colombia, Instituto Geofísico, Universidad Javeriana, Bogotá, Castellano e Inglés, 314p., 1977.

Sauer, W., Geología del Ecuador, Editorial del Ministerio de Educación, Quito, 383p., 1965.

Schaufalberber, P., Apuntes Geológicos y Pedológicos de la Zona Cafetera de Colombia, Tomo I, Federación Nacional de Cafeteros de Colombia, Imprenta Oficial Manizales, 295p., 1944.

Stauder, W., Tensional Character of Earthquake Foci Beneath Aleutian Trench with Relation to Sea-Floor Spreading, Jour.Geophys.Res., v.80, p. 7693-7701, 1968.

Stuader, W., Subduction of the Nazca Plate Under Peru as Evidenced by Focal Mechanisms and Seismicity. Jour.Geophys.Res., v. 80, p. 1053-1064, 1975.

Stutzer, O., Contribución a la Geología del Foso Cauca-Patía, Comp.Estud.Geol.Ofic. en Colombia, Tomo II, p. 69-140, 1934.

Tschopp, H.J., Geologische Skizze von Ekuador, Ver. Schweiz.Pet.-Geol. v.-Ing., Bull., v.15, no.48 p. 14-45, 1948.

van Andel, Tj.H., G.R.Heath, B.T.Malfait, D.F. Heinrichs, and J.I.Ewing, Tectonics of the Panama Basin, Eastern Equatorial Pacific, Geol.Soc. Amer.Bull., v. 82, p. 1489-1508, 1971.

Vergara y Velasco, F.J., Nueva Geografía de Colombia, Imprenta de Vapor, Bogotá, 1008p., 1901.

BOLIVIAN GEOLOGICAL GEOTRAVERSE

Salomón Rivas

Corporación Minera de Bolivia, La Paz

Abstract. In Bolivia new ideas have emerged concerning metamorphism, vulcanism and metallogenesis, so that the country may be considered a natural laboratory, giving rise to new hypotheses; an important factor for that is the strong erosion in the Cordilleras. Here the main aspects are introduced.

The Brasilian Shield, the Chapare window and the Cuprita debris (Northern Altiplano) have abundant metamorphic and igneous rocks derived from them; it may suggest the existence of ancient cordilleras now buried transverse to the Andes. It may be assumed that such rocks have been on surface, but now are deeply buried so that present outcrops are the summits of ancient cordilleras.

The volcanism is surficial, so that volcanic foci are 5 to 10 km deep in the Cordillera Occidental. In the Southern Altiplano (Lípez), volcanic are built on the ignimbritic Plio-Pleistocene plateau.

In the Cordillera Oriental the erosion has been so severe that volcanic foci are showing their roots, many of them corresponding to the igneous bodies containing important deposits of tin at the center and south of Bolivia.

In the light of the metamorphic and volcanic concepts mentioned above, a new metallogenesis theory is in investigation in the high Andes.

Geomorphotectonics

The Bolivian geotraverse introduced in this paper cuts all the important geological units of the country.

To the east the Brasilian Shield and neighboring lands are low flats, where gentle contrasts of relief differentiate the Brasilian Shield, Chiquitos Ridge and the Beni-Chaco Flat.

The Andean System begins to the west of those flats. This last moving part of the earth's crust extends parallel to the line of the Pacific coast. It begins with the Subsandino band which is strongly folded and constitute the main source of Bolivian oil.

The next geotectonic unit towards the west, is the Paleozoic block. This is the eastern Andean province, where Bolivian Paleozoic, Mesozoic and Cenozoic rocks are represented. In other parts of the country the same sequence is found and except for small facies changes, differences in thickness and the lack or addition of units, is stratigraphically similar.

Continuing to the west, the Altiplano unit is made up of continental sediments typical of intermountain basins and finally, the Cordillera Occidental chain is the mostwestern part of Bolivia, bordering with Chile, and is constructed of volcanics of Upper Tertiary and Quaternary age.

Tectonic Development

Bolivia has an important flexure of Paleozoic and Tertiary folds. Mountain chains enter Bolivia from Peru trending SE, but at the latitude of our geotraverse (Arica-Santa Cruz they inflect to trend NS and maintain this direction into Argentina. The major axes of folds, the main fault traces, and the outline of the igneous bodies (to which mineralized bands are related) follow those trends.

The Bolivian Geotraverse cuts all the morphostructural units; in it the three orogenic cycles identified in the country are represented, that is to say: Precambrian, Hercinian, and Andean.

Precambrian Orogensis. The latest investigations of the outcrops along the western border of the Brazilian Shield and in the east of northern Bolivia, show that Precambrian rocks, mostly metamorphic, are highly folded and the regional trend from east to west constracts with the folding of later cycles. This trend is maintained in the Chapare Precambrian window in Central part of our geotraverse and through the debris of gneiss and granite originating in the northern Altiplano. The regional lineation continues through the metamorphic rocks of Arequipa, Peru.

Recent studies show different Precambrian tectonic phases. These ancient rocks are only remnants of deeply eroded ancient mountains and conceal the magnitude of geological aspects of other better known cycles.

The radiometric ages of a few analyzed rocks vary between 1800 and 500 m.a.

Hercynian Orogenesis. Two stages of strong folding during the Paleozoic are well developed in the Cordillera Oriental and have influenced Andean Orogenesis.

The eohercynian tectonics ocurred during the Upper Devonian (Martínez et al., 1973). During that phase a large anticlinorium with a vertical axis was developed with assymmetrical folding and minor very tight and faulted refolding following the present trend of the Andes. The rocks near the anticlinorium core show schistosity and epimetamorphism.

The second Paleozoic folding, that is to say, the neohercinian, occurs during the Middle Permian, affecting Carboniferous sandstones and Lower Permian (Wolfcampian) limestones near the Peruvian border in the Copacabana peninsula. Structures formed in sedimentary layers are large and gentle but vertically uplifted and block faulted.

Andean Orogenesis. There are two folding models in two different logitudinal basins originating in central Peru: a) In the Subandean zone, where terrains from Paleozoic to Tertiary are almost concordant, they seem to be affected only by the Pliocene folding. Folds are assymmetrical and cut by many longitudinal inverse or low angle faults. b) In the Paleozoic block and Altiplano, the red beds from Cretaceous to Eocene age were folded. At the end of the Pliocene these were refolded together with volcanic and volcano-detritic terrains of Oligocene to Pliocene age; this movement was the most important in defining the structure and geography of the zone.

The tectonic configuration of major Andean structures mostly consists of symmetrical folds fractured into blocks by inverse and gravity faults dipping vertically.

Andean folding was followed by a continued denudation leaving mountains reduced to a low peneplain towards the end of the Miocene. That peneplain started to be uplifted by epeirogenic processes during the Pliocene. Uplift continued toward the Quaternary until reaching 6000 m. Then erosion was severe in the Paleozoic block, denuding metamorphic zones and volcanos to their roots and glaciers were formed on the higher remnants.

Rocks and Mineralization

Precambrian

The Precambrian appears in the Brasilian Shield (eastern part of fig. 2, Corresponding to Santa Cruz Department) where the most ancient and folded rocks are exposed. These are called generally "Basement Complex" and are poorly investigated, because they are covered by lateritic flats and hidden below a thick and hot forest with very limited access.

A large variety of rocks are found in the eastern Precambrian, mostly medium and high grade metamorphic: Schists, gneisses, migmatites, and anatectic granites, pegmatites and fairly long quartz veins cutting and crossing the area and containing indications of gold.

In the area called "Rincón del Tigre" gneisses, quartzites, ultramafic serpentinized rocks, gabbros, granites, and sandstones are folded in anticlines and synclines. Those rocks contain indications of chromium, nickel and copper mineralization.

In the alkaline zone of Velasco, not far north of our geotraverse, round bodies of syenites: normarquite, pulasquite, and foyaite lie on the highly folded and metamorphised complex; weathered carbonatite lavas (Fletcher, in Jordán, 1978).

Thick bands of quartzites and siltstones constiute the Sunsas Ridges, etc. (Oviedo, in Pareja, 1975).

In the Chapare Window anhydrites, gypsum and chalk outcrop diapirically, as well as limestones, dolomites and magnesites with some lutites, quartzites and marls which are quite folded. Asbestos and magnesite are promising resources to be developed.

In the Altiplano, gneissic and granitic boulders (Cuprita, Corocoro and Berenguela mines) are an indication that the complex Precambrian basement buried in the western end of geological section exhibited in fig. 1 is not deep. This corresponds to the eastern continuation of the Arequipa Massif. Copper is economically important in that altiplanic sector.

Cambrian

At the eastern end, red and violet sandstones, limestones and grey, green and black dolomites form the column of this system and lie discordantly on the Precambrian (Oviedo, in Pareja, 1975).

In the Chapare window Cambrian appears as a conglomerate of green matrix with all colour clasts, many of them igneous and metamorphic of hard texture.

No Cambrian rock outcrops in other sectors of our geotraverse, but they are present all along the cross section in the subsurface.

Ordovician

Ordovician is developed wonderfully in Bolivia in a broad band crossing central Bolivia from Peru to Argentina. The Cordillera Oriental shows the best and largest outcrops of this monotonous series of rocks difficult to differentiate.

In the Chiquitos Ridges there are arkosic and calcareous conglomerates and sandstones, then sandstones and siltstones with hematite forming the enormous iron deposit of Mutún. Some sandstones, siltstones and lutites constitute the higher part (Cabrera, in Pareja, 1975).

In the subandean central band some lutites and mostly quartzites and sandstones outcrop in several anticline cores (Canedo Reyes, 1960).

In the Cordillera Oriental, Cochabamba zone, the different stages are represented in a large

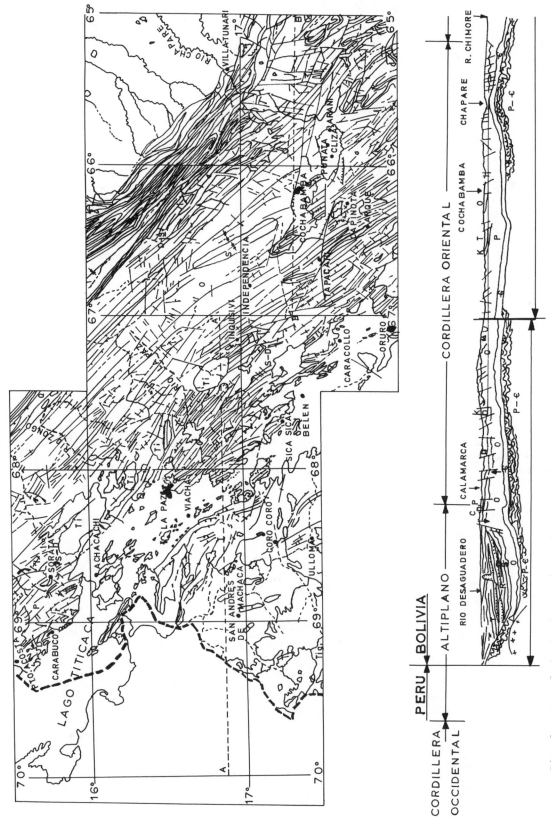

Fig. 1. Map of the band covering Bolivian geotraverse. Dashed lines AA' and BB' indicate section drawn below.

BOLIVIA   67

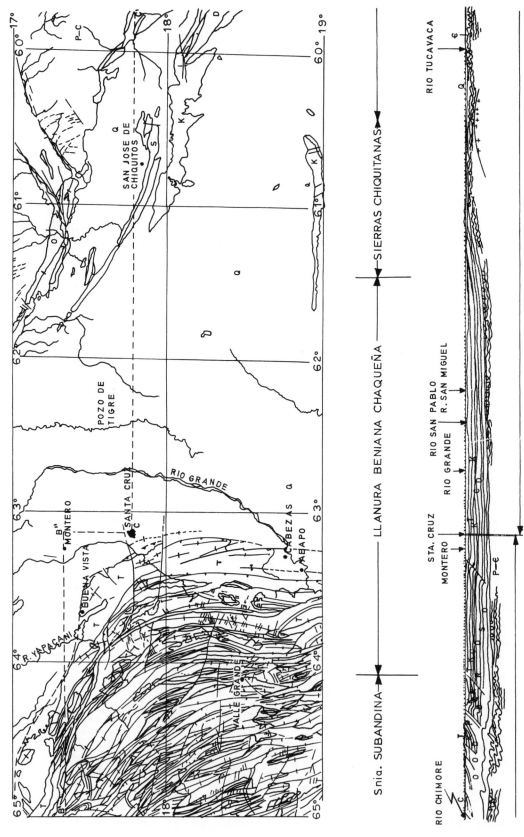

Fig. 2. Continuation of fig. 1 to the east (same trends as along CC' to 57.5°W not drawn). Dashed lines B'B" and CC' indicate section drawn below.

lithologic group. The lowest studied rocks are lutites and dark grey Arenigian siltstones. Llanvirnian-Llandelian sediments are largely distributed, being also lutites and dark grey siltstones. In turn almost horizontal Caradocian rocks are broadly scattered and made up of greenish grey to whitish sandstones and hard quartzites in thick layers (Rivas, 1971).

Mostly in southern Bolivia Ordovician rocks are important reservoirs of economical minerals; sandy levels are the wall rock for the antimony, lead and zinc veins.

Silurian

This begins with thin lutite and sandstone outcrops in the Chiquitos ridges, increases to moderate in the subandean band and then to thick development in the Cordillera Oriental, where it constitutes the wall rock of most important tin deposits of the country.

Spread over the Paleozoic block, or Cordillera Oriental, are translapping Lower to Upper Ordovician stages from south to north, the Wenlockian tillites, (diamictites) with a gentle discordance. A quartzite horizon follows, that is the lithologic control of tin mineralization. Immediately above is a thick, much fractured Ludlovian sequence of dark grey lutites with yellowish patina. Finally there is an upper sandy unit, being also an important lithologic control especially in the ore deposit of the Cordillera Tres Cruces (Koeberlin, in Deringer and Payne, 1937).

Devonian

The Cordillera Oriental, belonging to the Paleozoic block, has the largest development of Devonian rocks, as it did the rocks of the previous system.

Good outcrops may be seen along the La Paz-Oruro road and crossing the Cordillera towards Cochabamba, but its lower limit is not yet defined. Generally the basal Devonian is ascribed to a sandy crossbedded layer with ripple marks, red coloured, that follows to the last sandy brown Silurian unit.

This is followed by a thick pelitic lime-sandy sequence with many fossils including two to three sandy fossiliferous horizons and finally a sandy unit with few fossils (Fricke et al., 1964).

In the Subandean zone there are two sectors: the western one with units similar to those of the Cordillera Oriental (Vallegrande area); the eastern one (Santa Cruz area) where the lutitic series has very few intercalated sand lenses. Lutitic dark grey intercalations are abundant among the light grey micaceous and greenish sandstones at the top. These sandy levels are important reservoirs of Bolivian oil (Padula and Reyes, 1958).

At the base of the Chiquitos ridges a dark red conglomerate outcrops. Higher up is a whitish and reddish conglomeratic sandstone and a highly

fossiliferous horizon and above this a highly micaceous sandy grey yellowish unit together with dark grey micaceous fossiliferous lutites (Ahlfeld and Branisa, 1960).

In this way the main Paleozoic marine series, also acting as basement in the Altiplano, arrives at its end. Paleozoic rocks above are a sequence from the Carboniferous to the recent of predominately continental origin, but with many hiatus.

Carboniferous

The Subandean is rich in Carboniferous outcrops, but is only slightly developed in the Paleozoic block.

In the Subandean, the Carboniferous lies discordantly on different Devonian levels.

Its base is made up of diamictites (1); and grey to violet siltstones, then grey green sandstones, again dark grey diamictites interbedded with sandstones, conglomerates and lutites of same colour (2); in the last lutites there is lateral change to other sandstones and diamictites (3) of red colour. All these units constitute a basal group.

A massive series of yellowish and rose sandstone is discordantly superimposed with several conglomeratic lenticular horizons, ending with reddish siltstones and sandstones (Mather, 1922).

Several isolated ridges of the Chiquitos region are outcrops of carboniferous sandstones.

In the Andean block and Altiplano there are important Carboniferous outcrops about 50 km north and south of La Paz, that is to say, around Lake Titicaca on one side and in the Colquencha-Calamarca on the other side. All those rocks are pseudoconcordant on different Devonian levels.

In the Lake Titicaca zone, the Carboniferous is made up of arkosic whitish sandstone with remarkable crossbedding interbedded with grey greeny to dark grey sandstones. Above, whitish and violet sandstones culminate the sequence (Ascarrunz and Radelli, 1964).

In Colquencha and Calamarca there are good but isolated, outcrops, of rosy white sandstones or greenish greyish diamictites.

Permian

A marine transgression appears in the lower part of the sediments but the higher part of the sediments but the higher part of the column is again continental (Wolfcampian to Lower Leonardian).

In the Lake Titicaca region some layers of thin limestone outcrop. These are fossiliferous, dark grey, blue, green or red, and intercalated with some marls and lutites varying from light grey to black in colour.

Above that, we find discordant red arkosic sandstones, violet and brown; several pyroclastic levels and volcanic green, brown and purple effusions are intercalated (Cabrera and Petersen, 1936).

The Subandean Permian, the same as that antiplanic-andean is constituted mainly of fossiliferous limestones, some of them very silicified, with some intercalations of marl, dark grey lutites and some sandstones.

Triassic

In the Subandean zone the Triassic is represented by thin limestones with chert and flintstone nodules, whitish grey to dark grey; marls and clays are intercalated. At the base there is a thick yellowish rose-green sandy unit which is diagonally crossbedding.

At the top, locally and in southern Bolivia, there are anhydrites and lenticular bodies of rock salt together with clays, all of them red.

The Triassic lies discordantly on marine Carboniferous marine sediments.

Jurassic

No sedimentary rock is known in Bolivia from the Jurassic.

Cretaceous

There are Cretaceous outcrops in the Chiquitos ridges, in the subandean ridges in the Cordillera Oriental, and in the Altiplano. It was found through drilling in the Beni-Chaco plains.

Cretaceous rocks, remnants of two basins separated by the Marañon rise of Peru-Aiquile rise of Bolivia, lie on Paleozoic sediments with angular discordance. Generally speaking they are continental, lacustrian and paralic sediments, with short marine transgressions.

The Andean and Altiplanic Cretaceous, with some facies changes near the basin borders, begins with red brown conglomerates and sandstones from the Cenomanian, followed by multicolour limestones and marls, and ends in banks of quartzitic sandstones (Lohmann and Branisa, 1962).

In the Subandean zone and Chiquitos ridges the Cretaceous starts in some places with a basaltic flow of the Lower Senonian age, continues with strongly crossbedded red sandstones and ends with calcareous sandstones and sandy limestones with flintstone nodules. All the Chiquitos zone, called El Portón, is sandy.

Tertiary

The rocks of this system are largely disseminated in Bolivia with largest development in the western altiplanic region. In the rest of the Altiplano and in the Cordillera Oriental erosion has been severe.

Clastic sediments: sandstones, conglomerates and lutites, generally red beds, constitute the basal units transitional from the Cretaceous, and contain intercalations of gypsum layers indicating desert influence (Meyer and Murillo, 1961).

Conglomeratic sandstones and little consolidated light brown with many pyroclastic horizons consti-

ute the Tertiary upper units (Cheroni, 1975). Widespread layers of tuffs and ignimbrites to the west and of sandstone and clay brown grey to the east lie discordantly on previous folded sediments.

In that altiplanic region, the rocks are related to copper deposits and uranium traces.

In the Cordillera erosion has been severe so that only in tectonic trenches and adjacent to igneous bodies, such as those of Chocaya, Chorolque and Potosí, have continental and volcanic sediments been preserved. These are sandstones, conglomerates and red lutites with lavas and both light and dark tuffs intercalated (Turneaure and Marvin, 1947).

In the Subandean zone a basal conglomerate and sandstones is locally continued by siltstones and limestones with fossil residues. Thick layers of brown yellowish and reddish clayey sandstones continue the sequence with thick conglomerates containing tuff horizons near the top.

The Devonian to the Tertiary subandean rocks hold some promise for oil and gas.

In the Chaco-Beni flat along all the piedmont of subandean ridges Tertiary clayey sandstones constitute a zone of low hills. The gorges of major rivers cut massive clayey sandstones that are brown with varying yellow, grey and reddish tones.

In the Chiquitos ridges there are Tertiary sandstones and conglomeratic lenses with some limestones containing plant fossils (Almeida, 1945).

Quaternary

Again this system is represented in several places, but has the largest areal distribution and thickness in the Altiplano and the eastern plains. The irregular physiographic zones with their Tertiary sediments have been swept out by strong erosion.

The Quaternary has been formed through accumulation of deposits of lightly consolidated material.

In the Altiplano large zones covered by deposits of fluvio-lacustrian, alluvial, lacustrian, volcanic, glacial and eolian origin have been recognized.

Sands, gravels, clays, tillites, limestones, tuffs, ignimbrites, lavas are predominat accumulations. Finally there are salty lime deposits of this age: sodium, potassium, lithium chlorides and borax; and of peat, etc.

The Cordillera Oriental, bordering the Altiplano, contains La Paz Valley where the Quaternary of the glacial, interglacial and present fluvial is well exposed. Other isolated remnants of Quaternary deposition are around the towns of Cochabamba, Sucre, Tarija and in the Tiraque, Betanzos, Culpina, Tupiza and Villazon basins.

In the Subandean zone Quaternary remnants are preserved in large hanging synclines (Dobrovolny, 1956).

Recent alluvial deposits are found along the rivers; those below mineralized mountains form the tin placers, especifically in the Cordillera Oriental, Altiplano and high valleys. Similar collovial accumulations are on the ridge sides, in this case the tungsten earth (colluvial deposits) being important.

Almost the whole surface of the Chaco-Beni plains are covered by Quaternary sediments of alluvial, fluvial, lacustrian and lateritic soils, most of them covered by a luxuriant vegetation.

Igneous Rocks

Precambrian

The Brasilian Shield has large outcrops of anatectic granites, the result of high pressures and high temperatures in the Precambrian tectonic complexity. Recently, circular igneous cores with an east-west alignment located in the gneissic complex of the basement of nepheline syenite composition, syenitic quartz, together with carbonatite lavas have been mapped.

These igneous bodies, numerous metamorphic rocks, migmatites, pegmatites, granite and quartz dikes demand much more study.

Paleozoic

Little igneous activity of this age has been detected in Bolivia. Numerous dikes of diabasic-melaphiric composition fill certainly weakness planes in Ordovician and Devonian strata, sometimes with high carbonatite content.

Mesozoic-Cenozoic

In the Altiplano and both cordilleras igneous activity is important. The Cordillera Oriental has granodioritic batholiths aligned northwest to southeast, somewhat discordant related to the axes of Paleozoic basement folding; they seem installed on a peneplain. The strong erosion has left those hard resistant bodies as physiographic high, as well as their contact metamorphic rocks. The age of these igneous rocks may vary from the Triassic to the Tertiary; they are a good wall rock for veins, that is to say, a good lithologic control especially of tin and tungsten. However, the most important control for those minerals is found in small bodies outcropping near the batholiths of quartz porphyry composition extending to southern Bolivia. These have yielded valuable metals. In the fractures of cold rock those metals combined with oxygen, sulphur, etc. formed the veins being now exploited as the main asset of the Bolivian economy.

Stocks emplaced in Paleozoic rocks are found along lines in the Cordillera Oriental; their age may vary from a little before the Triassic, with greater quantities in the Tertiary (Pliocene-Pleistocene).

Cenozoic Volcanism

Large altiplanic areas and isolated areas of the Cordillera Oriental are covered with volcanic material; lavas, ignimbrites, tuffs and pyroclastics belonging to Plio-Pleistocene to Recent effusive phases.

The Morococala and Los Frailes-Livichuco high plateaus are well known, since in their windows important tin deposits have been discovered and it is assumed that volcanic material hides other deposits. The rocks are rhyolites and rhyodacites in composition.

In the Western Andes close to the Chilean border there are volcanic accumulations where cones, craters and lava flows still maintain their original forms. Going into Bolivia the craters are broken and eroded. Here the rocks are of composition dacitic and andesitic.

Some volcanic activity is present today in the Cordillera Occidental; it appears as fumaroles breaking the quietness of the clear sky of the desert.

References

Ahlfeld, F. and L. Branisa. Geología de Bolivia, Edit. Don Bosco, La Paz 215 p. 1960.

Almeida, F.F. Geología do Sudoeste Matogrossense, Div. Geol. Min., Brasil 116, 1-19. 1945.

Ascarrunz, R. and L. Radelli. Geología della peninsola di Copacabana e delle isole del settore sud del Lago Titikaka, Att. Sci. Natur., etc. 103(3); p. 273-284. 1964.

Cabrera-La Rosa, A. and G. Petersen. Reconocimiento geológico de los yacimientos petrolíferos del Departamento de Puno, Bol. Cuerpo Ing. Min. Perú. 115, p. 40-48. 1936.

Canedo Reyes, R. Informe sobre la Geología de la zona petrolífera del noroeste (reedición). Bol. Inst. Boliv. Petr. (IBP). 1(2). p. 9-31. 1960.

Cherroni, C. Geología del Distrito de Corocoro, Bol. Técn. YPFB.

Deringer, D.C. and J. Payne. The ore deposit of Llallagua in "Patiño leading producer of tin", Eng. & Min. Journ. 138, p. 171-177. 1937.

Dobrovolny, E. Geología del Valle Superior de La Paz, Bolivia, Bol Serv. Geol., La Paz. 1956.

Fricke, W.C. Samtleben, H. Schmidt-Kaller, H. Uribe and A. Voges. Geologische Untersuchugen im Zentralen Teoil des bolivianishchen Hochlandes nordwestlich Oruro, Geol. Jahrb., 83, p. 1-30. Hannover. 1964.

Jordán, G. Exploración de Minerales en el Oriente de Bolivia, Revista Ingeniería, Santa Cruz, p. 36-40. 1978.

Lohmann, H.H. and L. Branisa. Estratigrafía y y Paleontología del Grupo Puca en el sinclinal de Miraflores, Potosí, Petr. Boliv, 4(2), p. 9-16. 1962.

Martínez, C., P. Tomasí, T. Zubieta and R. Botello. Mapa Tectónico de Bolivia, Publ. Dept. Geosciencias, Facultad de Ciencias Puras y Naturales y Geobol, p. 1-4. 1973.

Mather, K.F. Front ranges of the Andes between Santa Cruz, Bolivia and Embarcación, Argentina, Bull. Geol. Soc. Amer., 33, p. 703-764. 1922.

Meyer, H.C. and J.E. Murillo. Investigaciones geológicas en la faja cuprífera altiplánica. Sobre la geología en las prov. Aroma, Pacajes y Carangas, Bol. Serv. Geol. Bolivia (Geobol), 1, 1961.

Padula, E.L. and F.C. Reyes. Contribución al Léxico Estratigráfico de las Sierras Subandinas, Bol.

Técn. YPFB, 1(1), p. 9-70. 1958.

Pareja, J. Discordancia en el extremo oriente de Bolivia, Rev. Técn. YPFB. 4,(3) p. 845-855.

Rivas, S. Ordovícico en el corazón de Bolivia, Geobol 15, p. 9-15. 1971.

Rivas, S. Nueva Teoría de Génesis de Yacimientos Minerales (inedited).

Turneaure, F.S. and T.C. Marvin. Notes preliminares sobre la Geología del Distrito de Potosí, Min. Boliv. 4(36), p. 9-14. 1947.

GEOPHYSICAL STUDIES IN CENTRAL ANDES

Ramon Cabré S.J.

Observatorio San Calixto, La Paz, Bolivia

Abstract. The Central Andes, especially in
Bolivia, are a very complex structure, being
involved with the curvature of the two cordilleras
and irregularities of the subducted Nazca plate.

In southern Bolivia there is some deep seismic
activity, a continuation of the Argentinian active
band, which is then truncated but continued on the
Peru - Bolivia border though displaced more than
500 km. Irregular Nazca plate geometry is analyzed through surface and intermediate seismic foci.

The bands of mineral deposits seem to be controlled by the underthrust moving plate; tin minerals probably originate in the continental mass,
some other minerals could be dragged down with the
oceanic plate.

The crust is 65 km thick beneath La Paz. Models
of wave velocity are introduced, according to P-
and S-residuals. P-Wave attenuation is high and
irregular.

Gravity anomalies indicate low density materials, possibly dragged from the Pacific bottom.

A band of high electric conductivity entering
from Peru extends south of La Paz and Cochabamba
and continues to the ESE.

Introduction

Bolivia is located in the heart of South America, traversed by the Andes. The north end of
the Andes sector trends NW-SE; in the center the
cordilleran elbow corresponds to that of the
coast at Arica; to the south the sector trends
N-S along the Argentina - Chile border.

The structure is extremely complicated both in
surface geology (Rivas, 1980) and in its deep
constitution. A double arc, divided by the
Quaternary fill of the Altiplano, probably corresponds to two phase changes in the materials
dragged down by the subducting plate whose dip
changes from north to south with irregularities
of folding and tearing.

It is mandatory in the present state of investigations to look for simplified models, which
in the future may be revised and improved.

Subducted Nazca Plate

The three dimensional distribution of shallow
and intermediate depth earthquakes corresponds

to the geometry of the Nazca plate (fig. 1).
Consequently relations between that plate and
tectonic, magmatic and mineralogic processes
may be found, at least for the last, say, 20
million years (Rodriguez et al., 1974; Minaya,
1978; Claure and Minaya, 1979).

It is well-known that in the Andes region there
is a remarkable gap of seismic activity between
350 km and 500 km depth. Moreover, in Bolivia,
deep seismic activity is absent except in the
south where some earthquakes originate at about

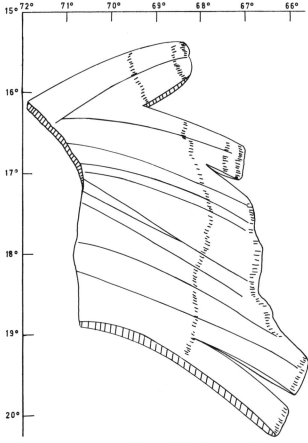

Fig. 1. Representation of the Nazca plate as
subducted beneath Central South America.

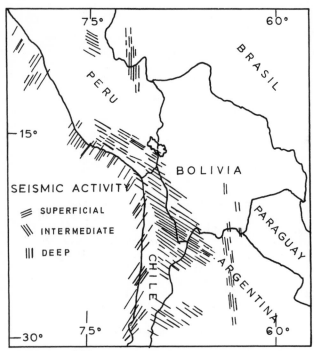

Fig. 2. Schematic indication of zones with
stronger seismic activity, according to focus
depth. Note the two parallel bands of deep
activity, which are displaced, leaving between
a gap filled with structural irregularities.

600 km as a continuation of a stronger deep
activity concentrated in Argentina along a band
across Santiago del Estero. That band is approximately parallel to Peru - Brazil border (fig. 2)
but displaced more than 500 km. (Such displacement is not caused apparently by any fault
movement but originated in the irregularities of
the subducted plate mentioned above).

From seismic activity, the subducted plate
appears to be 100 to 120 km thick, dipping gently
both in northern and southern Bolivia, more significantly and more irregularly at the center.

Among the irregularities, three wedges opening
towards the East were recognized as lacking
intermediate seismic activity. The northern one
encloses La Paz city and is characterized by the
presence of Mesozoic batholiths (120 to 210 m.a.)

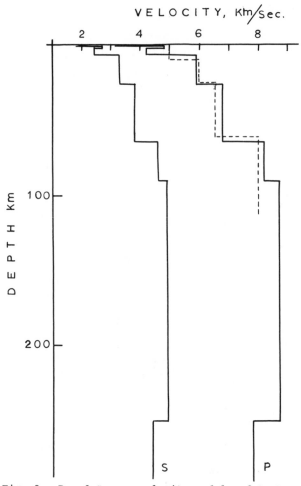

Fig. 3. P and S wave velocity models of La Paz,
for earthquakes in the Sandwich Is. (Solid lines,
Molina, 1977). P-Wave velocity Andean generalized
model (Dashed lines. DTM, 1970).

The second one, embracing the Cochabamba Valley
with rather high surface activity resembling a
graben, and with clear indications of strong tension on the Cordillera slope. Finally, the third
one filled 15 with young batholiths (9 to 20 m.a).

The Cordillera Occidental (Western Cordillera)
still maintains today some volcanic activity. The
Cordillera Oriental had some intervals of abun-

TABLE 1

P-Wave velocity generalized model of Andes and Altiplano (DTM, 1970).

Layer	Thickness, km	Velocity, km/sec.
1	10	5.0
2	15	6.0
3	35	6.6
4	. . . .	8.0

TABLE 2

P-Wave velocity models of La Paz and Peñas, for earthquakes of Nevada, U.S.A. (Medina, 1975).

	La Paz		Peñas	
Layer	Thickness, km	Veloc. km/sec.	Thickness, km	Veloc. km/sec.
1	0.4	3.2	0.2	3.2
2	2.0	4.8	2.0	4.8
3	5.0	4.2	2.0	5.8
4	18.5	5.9	9.5	5.9
5	39.1	6.8	52.0	6.8
6	55	8.1	55	8.1
7	180	8.6	180	8.6
8	>100	7.8	>100	7.8

dant volcanic effusions, the most recent being Tertiary.

Mineral deposits in the Bolivian Andes may indicate deep geodynamic activity. They are mostly distributed in bands trending parallel to the cordilleran arcs (also, indeed, there are some relations with lineaments observed on satellite images and with igneous bodies). This parallelism suggests that both orogenesis and metallogenesis are related to oceanic plate subduction. (Before the formation of the East Pacific Rise we should deal with the Pacific plate, but we may assume that its action against the Continent was not much different from present activity of the Nazca plate.

Tin minerals (the best known in Bolivia because of their economic importance) with bismuth and tungsten apparently have come up from the mantle, consequent to the deep disturbance produced by the subducted plate, though divergent opinions have to be recognized. Other minerals could be dragged down by the oceanic plate and then "distilled", or may originate from the continent upper mantle. In any event, water coming from the oceanic plate may have been the mineral solvent.

A very rough and preliminary estimation of temperature at the Benioff zone yields some correlation with formation of different minerals, though it has to be recognized that several deposits have been much diversified by a complex paragenesis.

Crust and Upper Mantle

Crustal thickness was found to be about 65 km beneath La Paz through the spectrum of longitudinal seismic waves (Fernández and Careaga, 1968). Several attempts to describe the Moho topography by means of explosion seismology have failed, because of the severe attenuation and a not well defined discontinuity (Aldrich et al., 1958).

Meanwhile it is assumed that the thickness is almost constant along the central cordilleran band and thins gradually on both sides.

A result coherent with this hypothesis may be seen in Vargas (1977). The 1978 explosion program of the Carnegie Institution of Washington led to the preliminary model of table 1.

More precise crustal and upper mantle models around La Paz (16°30'S, 68°08'W) were attempted through the inversion of both P and S residuals

TABLE 3

P- and S-wave Velocity of La Paz and Peñas, for earthquakes of Sandwich Is. (Molina, 1977).

	La Paz					
Layer	Thickness km	P-Veloc. km/sec.	S-Veloc. km/sec.	Thickness km	P-Veloc. km/sec.	S-Veloc. km/sec.
1	0.4	3.2	1.8	0.2	3.2	1.8
2	2.0	4.8	2.7	1.8	4.8	2.7
3	5.5	4.2	2.4	6.6	4.2	2.4
4	18.7	5.9	3.3	21.1	5.9	3.3
5	38.4	6.8	3.8	36.0	6.8	3.8
6	25.0	8.2	4.6	25.0	8.2	4.6
7	160	8.7	4.9	160	8.7	4.9
8	>100	7.8	4.4	>100	7.8	4.4

(table 2 and 3); fig. 3; see: Rodriguez and Medina, 1975; Molina, 1977).

Two low velocity layers are apparent: one thin and very irregular within the crust; the other in the upper mantle, beginning at a depth of 250 km. Otherwise wave velocity increases with depth, possibly by jumps. Remarkable variations are found between La Paz and Peñas (16°16'S, 68°28'W) which are quite close to each other. (tables 2 and 3).

P-Wave attenuation in the region is highly irregular so that attenuation appears negative (!) between Arequipa and La Paz for western foci and highly positive for the waves traveling the same way from east to west. Evidently such irregularity originates in the crust beneath both seismic stations, but it seems to be of regional character and not strictly local. Several studies have found a complex zonification of Q values in the region; see, for example, Sacks (1969).

Also wave velocities between La Paz and Arequipa are abnormal; Janes (1971a) finds high S-Velocity for an earthquakes originating in Central Africa; he finds difficulty in reconciling irregularities of Love and Rayleigh phase velocities.

The negative gravity anomaly in the Central Andes area is extremely high. This indicates low density materials, that to a large extent could be dragged from the Pacific bottom by the subducted plate to a hotter depth, so that they melted, segregated from denser rock, and then emerged to make the mountain roots (James, 1971a and 1971b; Cabré, 1974).

A band of high electric conductivity appears through anomalously large magnetic effects induced by rapid variations of the external magnetic field. This band enters northern Bolivia from Peru, curves to pass south of La Paz and Cochabamba and continues to the ESE (Aldrich et al., 1972). It is assumed to correspond to a high temperature band at low depth, without doubt related in some way to subducted Nazca plate.

References

Aldrich, L.T., H.E. Tatel, M.A. Tuve and G.W. Wetherill, The Earth's Crust, Carnegie Institution of Washington year Book 57, 104-111, 1958.

Aldrich, L.T., M. Casaverde, R. Salgueiro, J. Bannister, F. Volponi, S. del Pozo, L. Tamayo, L. Beach, D. Rubín, R. Quiroga and E. Triep, Electrical Conductivity Studies in the Andean Cordillera, Carnegie Institution of Washington Year Book 71, 317-320, 1972.

Cabré, R. Geodynamics in the Eastern Pacific and the Western Americas, Physics of Earth and Planetary Interiors, 9, 169-173, 1974.

Claure, H. and E. Minaya, Mineralización de los Andes Bolivianos an realción con la Placa de Nazca, Publicación de Programas ERTS, Observatorio San Calixto y COMIBOL, La Paz, 1979.

DTM Staff and Collaborators, Explosion Studies in the Altiplano, Carnegie Institution of Washington Year Book 68, 459-462, 1970.

Fernández, L.M. and J. Careaga, The Thickness of the Crust in Central United States and La Paz, Bolivia, from the Spectrum of Longitudinal Seismic Waves, BSSA 58, 711-741, 1968.

James, D.E., Anomalous Love Wave Phase Velocities, Carnegie Institution of Washington Year Book 69, 460-464, 1971a.

James, D.E., Plate Tectonic Model for the Evolution of Central Andes, Geol. Soc. Am. Bull. 82, 3325-3346, 1971b.

James, D.E., Andean Crustal and Upper Mantle Structure, J. Geophys. Res. 76, 3246-3271, 1971c.

Minaya E. Relaciones Fisico Quimicas de la Placa de Nazca y el Emplazamiento de Yacimientos Minerales, Thesis, Observatorio San Calixto, Public. 32, 1978.

Medina, A. Estructura debajo de los Andes a traces de Residuos y Atenuaciones de Ondas Sismicas, Thesis, La Paz, 1977.

Rodríguez, R., R. Cabré, and A. Mercado, Geometry of Nazca Plate Underneath South America and its Geodynamics Implications, Contribution to the SSA Meeting, March 24-31, 1974.

Rodríguez, R. and A. Medina, P-Time Residuals and the Low- Velocity Channel Underneath the Central Andes, Contributions to the SSA Meeting, March 24-31, 1974.

Rivas, S. Bolivian Geological Geotraverse, This Volume, 1980.

Sacks, I.S., Distribution of Absorption of Shear Waves in South America and its Tectonic Significance, Carnegie Institution of Washington Year Book 67, 339-344, 1969.

Vargas, F. 1-Discriminación 2-Cálculo del Espesor de la Corteza Terrestre Mediante Dispersión de Ondas Rayleigh Thesis, La Paz, 1977.

TECTONIC EVOLUTION OF PERUVIAN ANDES

Julio Caldas V. (Recopilation)

Instituto de Geología y Minería. Lima, Perú

Introduction

The Central Andes are considered to be a product of Nazca Plate subduction beneath the South American Plate. This interpretation is supported by the presence of a Benioff zone continuing landward from the oceanic trench and by the chronological progression of a magmatic and tectonic process from west to east (James, 1971), as a consequence of melting along that zone. However, Andean structures of the Peruvian continental margin show a thrust towards the east that does not explain the migration of South American Plate towards the west. Moreover, oceanographic investigations have not revealed any indication of compression, but rather of distention, which seems more consistent with deep block faulting through a long history of tectonism and magmatism (Myers, 1975b).

The nature of the Andean cycle is controversial. Ten years ago the literature was speaking about the Andean Geosyncline, a name later abandoned because the Andes do not have the components of the Alpine Geosyncline: ophiolites, syntectonic flysch, etc. Other classifications successively were proposed: intracratonic folded basin (Martínez et al., 1972), geoliminar basin (Aubouin and Borello, 1966; Aubouin, 1972) and finally geosyncline couple of a continental margin (Cobbing, 1976) in two parts, a western eugeosyncline and an eastern miogeosyncline. Independently of any theories, the Peruvian Andes are the result of the superposition of Precambrian, Hercynian and Andean orogens (Helwig, 1973) giving rise to an accumulation of 70 km of sialic crust beneath the Andes, one of the thickest in the world (James, 1971b).

The Instituto Geologico Minero y Metalurgico has produced the Tectonic Map of Peru, scale 1:2.500.000 updated to 1977 and a Tectonic Map of northwestern Peru, scale 1:500.000 (Agreement I.G.P. - INGEOMIN) with the following orogenic cycles (Fig. 1).

Orogenic Cycles

Precambrian Cycle

Precambrian rocks have been affected by undifferentiated polyphase tectonics mostly exposed along two bands of horsts, the one to the west constitut-ing the so called "Cordillera de la Costa" (Bellido and Narváez, 1960) the other towards the east the core of the Cordillera Oriental. According to recent investigations, the Precambrian basement is considered to be composed of two chronologically different series. The oldest one consists of granulitic gneisses and granites approximately 1800 m.a. in age (Cobbing et al., 1977) exposed mostly in the Cordillera de la Costa in southern Peru. The most recent rocks are pellitic, calcareous and quartzitic schists of green schist and amphibolite facies and in age approximately 540 m.a. (Cobbing et al., 1977), mostly exposed in the Cordillera Oriental and known as the Complejo del Marañón (Wilson and Reyes, 1964). Some remnants are found in the coast region (Cerro Illescas, Olmos, Marcona, Ocoña, etc.). The main exposures of Precambrian rocks trend from southwest to northeast.

Caledonian Cycle

In Peru there are also some probable remnants of a Caledonian cycle, since on the southern coast probable the existence of Siluro-Ordovician granitoids was proved. An example is the San Nicolás batholith, where a Rb/Sr determination in the total rock gave an age of 400 ± 22 m.a. (Ries, 1977), similar to that of some granites in the zone of Camaná. These, together with the absolute age of the gneisses, allows to assign to the Upper Precambrian and/or Cambrian a thick epimetamorphic pile exposed in the southern coast of Peru (Caldas, 1978).

Hercynian Cycle

The Hercynian chain is delineated approximately along the Cordillera Oriental. Remnants of Hercynian rocks have also been recognized in the coast zone (Cerro Amotapes and Illescas, Majes River, etc.). The Hercynian chain trends in general N 30°W. The folded Hercynian rocks are made up of abnormally thick sediments deposited during the Lower and Upper Paleozoic, amounting to between 10.000 and 15.000m (Martínez et al., 1972). The Lower Paleozoic lies discordantly on the Precambrian basement and is made up of pellitic or sandy

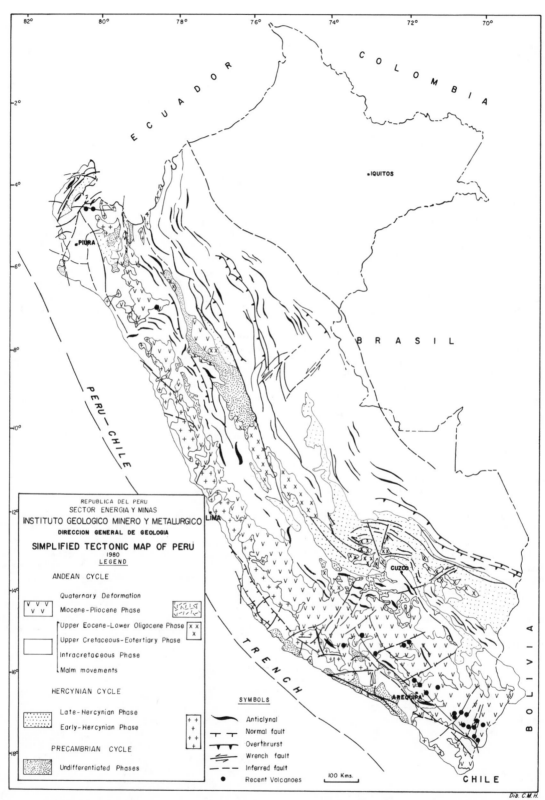

Fig. 1. Simplified Tectonic Map of Peru showing the regions deformed during Alpine, Hercynian and Precambrian orogenic cycles.

flysch type sediments while the Upper Paleozoic is made up mostly with continental molassic sediments overlain by marine deposits.

Hercynian tectonics resulted from two deformation phases, the first, or Eohercynian phase is well known in the Peruvian Andes (Megard et al., 1971; Martínez et al., 1972). It is characterized by the presence of polyphase structures; a primary folding of decimetric dimensions trending N 30°W with remarkable schistosity, crossed by a second folding,

Hercynian tectonics resulted from two deformation phases. The first, or Eohercynian phase is well known in the Peruvian Andes (Megard et al, 1971; Martínez et al, 1972). It is characterized by the presence of polyphase structures; a primary folding of decimetric dimensions trending N30°W with remarkable schistosity, crossed by a second folding, also decimetric in dimensions, trending N60°E, almost orthogonal to the first one. Associated with this tectogenesis is a regional metamorphism not more intense than green schist facies. It was accompanied by the emplacement of syntectonic granites with progressive metamorphism, recognized in the Amparaes dome (Marocco, 1978), at San Gabán (Laubacher, in press) and at the massif of Illescas (Caldas, in press). There are also examples of Eohercynian post-tectonic granites. The compressive deformation was followed by block faulting, preparing sedimentary basins for the Upper Paleozoic.

The second, or Late Hercynian, deformation varies in both nature and intensity in different parts of Peru. In the south is is characterized by a compression producing large folding with axial planes striking SW associated with an earlier stage of fracture schistosity. In the central region this phase consists only of a general uplift accompanied by block faulting trending NW-SE. In many cases these structures have been reactivated during the Andean cycle (Megard, 1978). An important plutonism, mostly of granitic composition was associated with this phase, examples being the granites of San Ramón, Machu Pichu, Coasa, etc. (Capdevila et al, 1977) A Rb/Sr determination in total rock indicates an age of 238±10 m.a. The deflections of Huancabamba and Abancay appear to have initiated their development during this phase (Marocco, 1978).

Andean Cycle

In the Mesozoic depositional stage, called by different authors geosynclinal, geoliminar, etc., the Andean basin in the Peruvian part has characteristically two bands of accumulation; a volcano-clastic western one almost 7000 m thick, and a clastic eastern one with 6000 m of sediments controlled by the presence of faulted basement blocks parallel to the western margin. In terms of plate tectonics, subduction could originate the formation of negative areas acting as volcanic or clastic sedimentation centers, controlled by significant crustal shortening and followed presently by crustal thickening to almost 70

km, among the thickest in the world (James, 1971). The details of this process are not yet clearly understood (Cobbing, 1976).

In the Late Cretaceous and beginning Cenozoic, the Andean axial band began to be generally uplifted and in the Tertiary, intermittently, some strong subaerial volcanic activity developed. The uplift gave rise to a remarkable subsidence through gravitational faulting in the coastal and subandean regions. In the former area, faulting controlled the formation of marine basins on the platform; in the latter, it controlled the continuation of marine sedimentation, which progressively changed to continental.

Movements in the Upper Jurassic, ascribed to the Nevadian deformation, were epeirogenic in nature, and affected a large part of Peruvian territory. Their effects are seen in the erosional discordance between the Dogger and the Malm and in continental redbed deposition, mostly in the subandean region, derived from erosion of uplifted blocks, among them the Marañón geoanticline.

Intra-Cretaceous deformation is limited essentially to the Pacific side of Peru and affected the Upper and Middle Albian volacno-clastic marine series, identified in northern and central Peru as the Casma group. According to Myers, (1975a), this deformation is characterized by intermittent pulses during the intrusion of gabbros that were precursors of the central batholith. In the Middle to Upper Albian its action resulted in folding of the Mesozoic series into large folds with vertical axial planes. On the other hand, the largest tonalite-granodioritic phases of the coastal batholith intrude previously folded rocks without causing any structural deformation of the wall rocks, but originating in these rocks a light regional metamorphism of the zeolite facies or, in the deepest levels, of green schist and amphibolite facies.

The coastal batholith has been demonstrated to be a multiple intrusion with a sequence from basic to acidic emplacing gabbros, tolalites and granites in a band 1000 km long in the core of the Cordillera Occidental (Pitcher, 1977). The rocks of the batholith may be differentiated according to plutonic units with related units forming super-units with characteristics of compositional segmentation. Following this criterion, three segments of the batholith have been recognized, Trujillo, Lima and Araquipa, that may correspond to that may correspond to structural and metallogenic segmentation of the Andes related to variations in dip of the Benioff zone (Cobbing et al, 1977). The Lima segment is the best known and is characterized by plutonism extending from 100-95 m.a. to 30 m.a. During this time the volcanism developed, ascending in many cases to the surface along annular complexes, depending on the physical nature of the magmas. These annular conplexes indicate that the volcanism was developed by subsidence of calderas, which at the surface were the foci of volcanic emission (Myers, 1975a; Bussell et al, 1976).

Steinmann (1929) called the end Cretaceous-Eo-tertiary deformation the "Peruvian phase". It corresponds to a general uplift of the Andes to a height of 1000 m above sea level during a broad and large folding, causing a complete withdrawal of the sea from the present cordilleran region. The molassic deposits of this phase are known in Peru as the Chota, Huaylas, Casapalca and Puno group formations.

The Upper Eocene-Lower Oligocene phase was called by Steinmann the "Incaic plase". It is responsible for the largest compression of the Andean cycle, expressed by intense folding with super-imposed effects, mostly in the cordilleran region, controlled by the lithologic nature of the affected formations. For example, the lutites of the Chicama formation or evaporitic sequences of southern Peru acted as lubricants. The deformation involved the Precambrian basement, evident in the inverse faulting visible in the coastal region and in the Marañón geanticline where examples are found of these rocks superposed or imbricated with Mesozoic sequences.

As a consequence of that compression, the sialic crust of the Peruvian Andes thickened considerably accompanied by remarkable transverse deformation in the deflections of Huancabamba and Abancay, modifying previous structures. After that, intense magmatic effusive activity developed, associated with plutonism.

In the Miocene-Pliocene phase, called by Steinmann (1929) the "Quechua phase", Miocene continental sediments were folded and then discordantly covered by Lower Pliocene sediments. Recent work carried on mostly in central and southern Peru (Megard, 1978) revealed the existence of two folding phases affecting post-Eocene deposits. The Oligo-Miocene material is affected by two deformation subphases, an intra-Miocene subphase affecting the volcanic series with a minimum K/Ar age of approximately 20 m.a. (Noble et al, 1972), and a Mio-Pliocene subphase affecting a tuff series, mostly ignimbritic, of K/Ar age approximately 14 m.a. The Mesozoic-Cenozoic sedimentary sequence of the subandean band was compressed for the first time in the Mio-Pliocene phase.

After that tectonic phase, epeirogenic movements uplifted the Andes at least another 3000 m, originating remarkable longitudinal faulting and giving occasion to Pliocene-Quaternary volcanism, such as the Barroso volcanic arc in southern Peru (Mendivil, 1965). This arc is parallel to the Arica trench revealing a close connection with the morphology of the Benioff zone.

Quaternary volcanism was characterized by an opposite tendency to the magmatism developing from the Triassic to the Miocene. During the latter time, migration of magmatism was from west to east, while in the Quaternary, magmatism migration was from east to west. This raises the question whether this magmatism still belongs to the Andean cycle or rather that we are dealing with a new geologic process.

Quaternary deformation is characterized essentially by fracture tectonics, mostly normal faulting, though in the Mantaro Valley there is evidence of lacustrine materials being lightly folded and flexured (Megard, 1978). Quaternary faulting, with reactivation of existing faults, happened during the 1946 and 1969 earthquakes in Callejon de Conchucos and Huaytapallana respectively. A classic example of recent faulting is in the Valle de Volcanes of Andahua in southern Peru. During this process several Pliocene-Quaternary volcanoes have also been reactivated. In the coastal region, Quaternary deformation is characterized by uplift and formation of "tablazos", or marine terraces, with fracturing mostly along large pre-existing faults.

References

Aubouin, J., Chaines liminaires (Andines) et chaines geosynclinales (Alpines), 24th IGE, Section 3, 1972.

Aubouin, J., and A.V.Borello, Chaines andines et chaines aplines: regard sur la geologie de la Cordillerre des Andes au parallele de l'Argentine moyenne, Bull. Geol. Soc. France 7, VIII, 1966.

Bellido, E., and S. Narváez, Geología del cuadrángulo de Atico, Co. Carta Geol. Nac. Bol. 2, 1960.

Bussell, M.A., Fracture control of high level plutonic contacts in the Coastal Batholith of Peru, Proc. Geol. Ass. 87, 237-246, 1976

Caldas, J., Geología de los cuadrángulos de San Juan, Acari y Yauca, Inst. Geol. y Min. Bol. 30, 1978.

Caldas, J., El complejo metamórfica de Illescas, IV Congreso Peruano de Geología (en prensa), Bol. Soc. Geol. de Perú.

Capdevila, R., F. Megard, J. Paredes, and F. Vidal, Le Batholite de San Ramon, Cordillere Oriental de Pérou Central, Geol. Rundschau, Stuttgart, 1977.

Cobbing, J., The geosynclinal pair at the continental margin of Peru, Elsevier Scieitific Publishing Co., Amsterdam, 1976.

Cobbing, J., J.M. Ozard, and N.S. Snelling, Reconnaissance geochronology of the crystalline basement rocks of the Coastal Cordillera of southern Peru, Bull. Geol. Soc. Amer. 88, p. 241-246, 1977.

Helwig, J., Plate tectonic model for the evolution of the Central Andes; Discussion, Bull. Geol. Soc. Amer. 84, p. 1493-1496, 1973.

James, D., Plate tectonic model for the evoluion of the Central Andes, Bull. Geol. Soc. Amer. 82, p. 3325-3346, 1971.

Lancelot, J.R., G. Laubacher, R. Marocco and U. Renaud, U/Pb radiochronology of two granitic plutons from Eastern Cordillera (Peru), Geol. Rundschau, Stuttgart, 1978.

Laubacher, G., Geología de la región norte del Lago Titicaca, Inst. Geol. Min. y Met., Bol. 5, Serie A, 1979.

Marocco, R., Estudio geológico de la Cordillera

de Vilcabamba, Bol. Inst. Geol. y Min. No. 4, Series A, 1978.

Martínez, C., P. Tomasi, B. Dalmayrac, G. Laubacher and R. Marocco, Carateres generaux des orogenes precambriens, hercyniens et andins au Perou et en Bolivie, 24th Int. Geol. Cong. , Sec. 1, 1972.

Megard, F., B. Dalmayrac, G. Laubacher, R. Marocco, C. Martínez, J. Parades, and P. Tomasi, La chaine hercynienne au Pérou et en Bolivie, premiers resultats, Cah. ORSTOM, Ser. Geol. III, 1971.

Megard, F., Etude geologique des Andes du Pérou, Central Service des Publications, ORSTOM, Paris, 1978.

Mendivil, S., Geología de los cuadrangulos de Maure y Antajave, Bl. Com. Carta. Geol. Nac. 10, 1965.

Myers, J., Cauldron subsidence and fluidization: mechanisms of intrusion of the Coastal Batholith of Peru into its own volcanic ejecta, Bull. Geol. Soc. Amer. 86, p. 1209-1220, 1975a.

Myers, J., Vertical crustal movements of the Andes in Peru. Nature, 254, No. 5502, 1975b.

Noble, D.C., E.H. McKee, E. Farrar and U. Peterson, Episodic Cenozoic volcanism and tectonism in the Andes of Peru, Earth and Plan. Sci, Let., 21, 1972.

Pitcher, W.S., The anatomy of a batholith, Jour. Geol. Soc. Lon. V. 135, part 2, 1978.

Ries, A.C., Rb/Sr ages from the Arequipa Massif, southern Peru, 2oth Ann. Rep., Res. Inst. Af. Geol. Univ. Leeds, p. 74-77, 1977.

Steinmann, G., Geología del Peru, Univ. Heidelburg, 1929.

Wilson, J., and L. Reyes, Geología del cuadrángulo de Pataz, Bol. Com. Carta Geol. Nac. No. 9, 1964.

NAZCA PLATE AND ANDEAN FOREARC STUDIES

L.D. Kulm, J. Dymond, and K.F. Scheidegger

School of Oceanography, Oregon State University, Corvallis, Oregon 97331

Abstract. The Nazca lithospheric plate was
studied extensively during the decade of the
1970's. Primary emphasis was placed upon the
East Pacific Rise (EPR) and the Andean convergence zone. The most recent spreading rates on
the EPR are variable and average about 8 cm/yr;
spreading is asymmetrical in some areas. The EPR
is characterized by different morphologies related to its spreading history.

Older crust generated by the fossil Galapagos
Rise displays four layers with a subcrustal layer
(7.2 km/sec) between layer 3 and the mantle. The
Nazca plate crustal section has a generally higher velocity and the upper lithosphere has a
greater mass than other Pacific sections. This
oceanic slab is deforming within the Peru-Chile
trench by extensional and compressional faulting;
it is also offset by large tear faults. The slab
is traced about 60 km beneath the Andean forearc
off Peru. The nature of the overlying subduction
complex is unknown but numerous forearc basins
off Peru contain Cenozoic terrigenous clastic
sediments with late Cenozoic dolomitic limestone
being prominent in the seaward basins. Numerous
tectonic events have disturbed the basins and
underlying complex.

Upper oceanic layer 2 basalts recovered from
the interior and margins of the Nazca plate are
characteristically more fractionated than their
counterparts from the Atlantic Ocean basin. It
is postulated that large, steady-state magma
chambers beneath fast spreading centers facili-
tate mixing of primitive magma batches derived
from the underlying mantle with much larger crustal level reservoirs of differentiated magma.
Mantle plume activity appears confined to Easter
Island and associated seamounts and the Galapagos
Islands.

Metalliferous sediments are ubiquitous on the
spreading East Pacific Rise and in the adjacent
Bauer Basin. They were formed by precipitation
from hydrothermal solutions generated by the interaction of seawater with the newly formed crust.
Chemical and isotopic data indicate both mantle
and seawater sources of elements are important in
the metalliferous sediments.

Introduction

The Nazca plate is an intermediate-sized lithospheric plate which is bounded on three sides by
spreading centers: East Pacific Rise, Galapagos
Rift, and Chile Ridge, and on one side by the
Peru-Chile trench (Figure 1). It occupies approximately 16×10^6 km^2 of the southeast Pacific
Ocean. Previous studies of the major features
with positive relief within the plate include the
fossil Galapagos Rise, Sala-y-Gomez Ridge and the
aseismic Nazca Ridge [Mammerickx et al., 1975].
The Bauer Basin and Yupanqui Basin immediately
to the east of the East Pacific Rise are the major features of negative relief on the plate.
Several rather low relief fracture zones traverse
the plate and extend toward the trench and continental margin.

The Nazca plate converges with the South American plate at a rate of 10 cm/yr, producing the
Peru-Chile trench. It displays the greatest
depth (8065 m) in the eastern Pacific Ocean and
one of the most active inclined seismic zones in
the world [Kelleher, 1973]. The associated Andes
Mountains constitute the longest orogenic belt
in the world with extensive volcanism and plutonism [Ericksen, 1976]. Mineral deposits are
found throughout the Andes, but are more numerous in the latitudes between northern Peru and
central Chile [Fig. 1; Ericksen, 1976].

Several studies in the decade of the 1970's
have described the evolution of the Nazca plate
throughout its Cenozoic history [e.g., Herron,
1972; Rea and Malfait, 1974; Handschumacher,
1976; Mammerickx et al., 1975, 1980]. This evolution began with the spreading history of the
now extinct Galapagos Rise in the center of the
plate and ended with the relatively recent
spreading history of the East Pacific Rise.
Earlier studies recognized the high concentrations of transition metals in the sediments covering the East Pacific [Bostrom and Peterson,
1966, 1969] which led to the more extensive studies of the so-called metalliferous deposits associated with the rapidly spreading (8 cm/yr) East

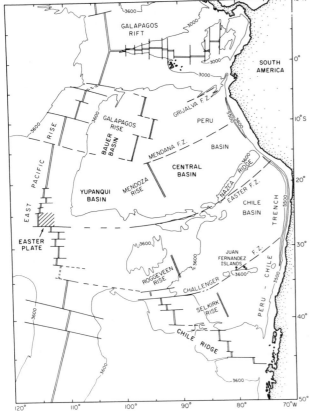

Fig. 1. Physiographic features of the Nazca
plate. Modified from Mammerickx et al., 1980.

the second part of the report, we propose new
areas of investigation that will further eluci-
date the crustal evolution of the Nazca plate and
the effect of its convergence with the South
American continental block.

Part I: Crustal Structure, Composition and Tectonics of East Pacific Rise and Nazca Plate

Evolution of the East Pacific Rise

Numerous detailed studies [Rea, 1981, 1979,
1978, 1978a, 1976, 1978a, 1975; Rea and Blakely,
1975] of selected portions, and reconnaissance
surveys of other portions of the East Pacific
Rise, show the shallow structure and tectonic
framework of a fast spreading mid-ocean ridge
(Figure 1). Five regions along the Nazca-
Pacific plate boundary were the sites of detailed surveys completed during the Nazca plate
project. Four surveys were located in relatively typical axial regions and one in the area of
the tectonically complex Easter plate. Reconnaissance tracklines also cross the spreading
center. Sea-floor spreading occurs at about
160 mm/yr (whole rate) along that portion of the
East Pacific Rise studied [Rea, 1981]. South of
the fracture zone at 13.5°S, spreading is
asymmetrical and faster to the east (Figure 1).
Spreading rates have not been constant along the
rise but rather have varied through time. They
decreased from at least 3.8 m.y. ago to a minimum during the Jaramillo to Olduvai magnetic
interval centered 1.32 m.y. ago. They have
increased about 7% to their present values since
that time.

Rea [1981] notes that the lithosphere near
fast-spreading rises is quite thin and is probably more susceptible to deformation than along
slow-spreading ridges (e.g. mid-Atlantic Ridge).
The effects of weak lithosphere may take several
forms: the disordered, blocky morphology of the
active portion of the 9°S fracture zone; the
presence of twinned spreading centers surrounding the small Easter plate; and the existence of
five small, 10-15 km, offsets of the spreading
center. Such offsets all were formed within
about 0.5 m.y. and affect an axial length of
100-200 km. It is postulated that they are the
result of small, discrete axis jumps facilitated
by the unusually thin lithosphere [Rea, 1981].
The history of formation of the East Pacific
Rise between the 13.5°S and 4.5°S fracture zones
(Figure 1) is determined from reconnaissance
data in this region. Spreading activity shifted
600-850 km westward from the old Galapagos Rise
which is now located in the center of the Nazca
plate [Mammerickx et al., 1980], to its present
location by three large jumps. Each of these
jumps resulted in the formation of a fracture-
zone bound section of the new rise. The ridge

Pacific Rise and adjacent Bauer Basin [Dymond
et al., 1973].

The first large-scale investigation, using explosion seismology, of the crustal structure of
the Nazca plate was conducted by Hussong et al.
[1976] and showed the presence of a subcrustal
layer (7.4 km/sec) overlying the upper mantle.

Pioneering studies by Zeigler and others [1957],
Risher and Raitt [1962], Scholl et al., [1968,
1970], and Hayes [1966] described the morphology
and crustal structure of the Peru-Chile trench
and adjacent Andean continental margin. Later
studies showed the extensive and rapid deformation occurring within the trench [Kulm et al.,
1973; Prince and Kulm, 1975; Schweller and Kulm,
1978]. More recent investigations show great
variations in the crustal structure, distribution of sedimentary basins and overall evolution
of the continental margin [Hussong et al., 1976;
Masias, 1976; Coulbourn and Moberly, 1977; Getts,
1975; Kulm et al., 1977; Whitsett, 1976, and
Shepherd, 1979].

In the first part of this report we present an
overview of some of the most recent large-scale
studies and summary works just completed on the
Nazca plate and adjacent continental margin.
They are being prepared for publication in the

jumping process covered about 2.5 m.y. (i.e., from 8.2 to 5.7 m.y. ago) [Rea, 1981].

Crustal Structure of Nazca Plate

Seismic refraction studies of the Nazca plate oceanic crust [Hussong et al., 1976] show that it varies in thickness from about 10 to 13 km, and displays at least four layers with a ubiquitous subcrustal layer of velocity 7.2 km/sec. In addition, drilling on DSDP Leg 34 [Yeats, Hart et al., 1976] indicates that layer 2 on the eastern edge of the plate is highly fractured in the upper few tens of meters. The composite Nazca plate layering sequence and values are given in Table 1 [after Hussong et al., 1976]. The data from this study suggest that the typical Nazca plate crust is shallow and thin (for a given age) and composed of high-velocity material. This crustal character may be related to the very high sea-floor spreading rates on the plate (about 8 cm/yr). When the composite section shown above is compared with standard Pacific Ocean sections [e.g., Worzel et al., 1974; Hussong et al., 1976; Shor et al., 1970], it appears that the Nazca plate crustal section has a generally higher velocity, due largely to the thick 7.2 km/sec basal crustal layer. Furthermore, the mass of the Nazca plate upper lithosphere is greater than the mass of other Pacific Ocean crustal sections. It appears that at a given age the denser upper lithosphere is associated with faster spreading such as occurs on the Nazca plate [Hussong et al., 1976].

Composition of Nazca Plate Basalts

Spreading Centers and Fracture Zones

Using basalts from 38 locations on the East Pacific Rise, 18 locations on the Galapagos spreading center, 11 locations on the Easter fracture zone (hotline) and 21 other locations on the Nazca plate from crust believed to have formed at the now extinct Galapagos Rise, Scheidegger and Corliss [1981] studied the crustal rocks of the Nazca plate (Figure 2). They have found that the compositional variations among Nazca plate basalts can be best described in terms of five populations of samples: southern East Pacific Rise (SEPR; 38°-23°S); northern East Pacific Rise (NEPR; 23°S-3°N); Easter Island and associated seamounts and islands of the Easter "hot line"(EI), Galapagos spreading center (GSC), and central and eastern Nazca plate samples associated with the extinct Galapagos Rise (GR). The Galapagos spreading center and Easter Island populations exhibit little compositional overlap. The EI basalts are rich in alkali metals, show moderate iron enrichment, but do not exhibit a typical Skaergaard trend as do GSC samples. All EI basalts show strong light rare earth element (REE) fractionation, whereas, REE patterns from

GSC exhibit a complete range from typical mid-ocean ridge basalt (MORB) patterns to moderate LREE enrichments. Even though both EI and GSC are thought to be influenced by mantle plume activity, they bear little compositional similarity.

The northern East Pacific Rise, southern East Pacific Rise, and the Galapagos Rise populations (Figure 2) are dominated by basalts with typical MORB elemental abundances, including LREE depleted patterns, and familiar major element variations [Scheidegger and Corliss, 1981]. Significant compositional differences exist between the SEPR and NEPR populations even though fractionation-induced variability partially obscures the regional trends. For example, the NEPR basalts represent much more primitive parental magmas (lower FeO*/MgO, lower concentration of total REEs, higher transition metal abundance) than the SEPR basalts. These investigators further note that the NEPR form a rather bimodal population, being either quite primitive or more highly fractionated, whereas, the SEPR basalts are much more uniform and occupy a compositional field between the two extremes of the NEPR population. If both East Pacific Rise populations are compared with the Galapagos Rise Population, one finds again close overall compositional similarity, particularly for the unaltered samples (defined as $Fe_2O_3/FeO < 0.5$) from GR. Lastly, basalts recovered from the Galapagos spreading center are strikingly similar to those from NEPR population, both in regard to major element trends, REE patterns, and other trace element abundances.

The several factors that are thought to be important in controlling the composition of sea floor basalts on the Nazca plate include: (1) low pressure fractional crystallization involving plagioclase, clinopyroxene and olivine; (2) mantle inhomogeneities; (3) low temperature oxidative alteration; and (4) rapid seafloor spreading [Scheidegger and Corliss, 1981]. Low pressure fractional crystallization is thought to be primarily responsible for the large chemical variations noted among basalts recovered from particular areas. Fractionation appears to impart considerable compositional "noise" that tends to mask possible regionally significant trends. Differences in the composition of mantle source rocks (mantle inhomogeneities) may

TABLE 1

Layer	Velocity (km/sec)	Thickness (km)	Density (gm/cm^3)
1	1.7	0.2	1.78
2	5.6	1.1	2.62
3	6.4	1.5	2.79
subcrustal	7.2	3.6	3.00
mantle	8.3	-	3.37

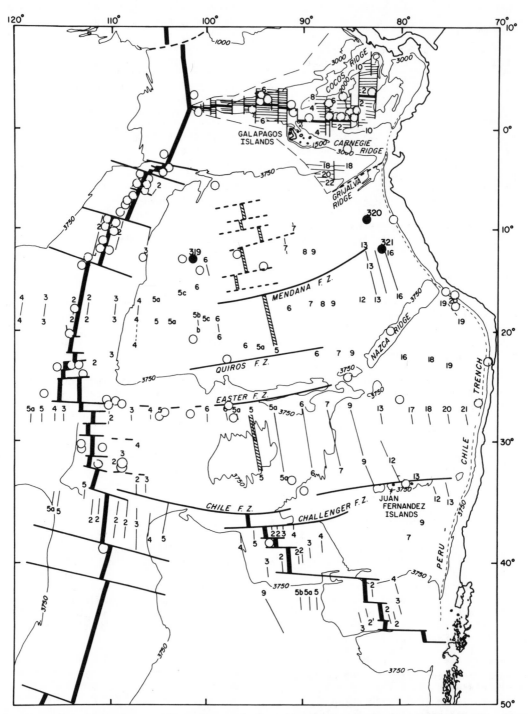

Fig. 2. Location of dredge samples of oceanic basalts (open circles) and DSDP drill sites 319, 320 and 321 (solid circles). Numbered magnetic lineations and structure from Herron[1972] and Handschumacher [1976].

be responsible for noted differences in rare earth element patterns between the Easter hot line and other populations and between the southern and northern East Pacific Rise. Mantle plume activity appears responsible for bringing compositionally different source rocks into the magma generation zone beneath the Easter hot line, but the importance of such activity for the GSC is less clear. Basalts dredged from the NEPR where mantle plume activity is not suspected are compositionally quite similar to those associated with the GSC where such activity is believed important. Basalts on the EPR and GSC are remarkably fresh, whereas basalts recovered from the interior of the plate and on topographic highs are highly altered and show the expected compositional effects of alkali and iron enrichment and magnesium depletion. Prominent fracture zone offsets of the GSC may be compositional interfaces between opposing segments of the spreading center, but similar offsets along the EPR do not appear to influence basalt compositions. Continuous, not discontinuous, compositional changes occur along the East Pacific Rise. Basalts recovered from the interior and margins of the Nazca plate are generally more evolved (higher FeO^*/MgO ratios and TiO_2 abundances, lower CaO and Al_2O_3 concentrations) than their counterparts from the slow spreading Mid Atlantic Ridge. Scheidegger and Corliss [1981] suggest that such differences may correspond to fundamental differences in thermal regime beneath fast and slow spreading centers and to the size and continuity of shallow magma chambers. Specifically, these workers suggest that large continuous magma chambers beneath fast spreading centers of the Nazca plate favor slow cooling and pronounced fractionation, leading to a preponderance of iron-rich basalts.

Subducting slab

The composition of the oceanic crust entering the subduction zone has been determined from exposures of basalts dredged in the vicinity of the Peru-Chile Trench between 9° and 27°S latitude [Scheidegger et al., 1978] and in the three DSDP holes (site 319, 320 and 321) drilled in the Nazca plate and described in Yeats, Hart et al., [1976] (see Figure 2).

Abundant pillow basalts dominate the rocks dredged from the trench [Scheidegger et al., 1978]. They are olivine and quartz-normative tholeiites that are believed to have formed at the now extinct Galapagos Rise 30-50 m.y. ago. Detailed chemical analyses of the basalts indicate that considerable compositional variability exists both within each of the dredged areas as well as between areas. Most of the inherent chemical variability observed within particular basement sections appears consistent with the concept of temporal evolution of magma bodies at a former fast spreading center by extensive shallow level fractional crystal-lization involving primarily plagioclase and olivine. In contrast, important chemical differences between the dredged areas suggest compositional heterogeneities in the mantle source regions.

Scheidegger and Stakes [1977] have studied the nature of secondary alteration of basalts exposed on fault scarps in the Peru-Chile Trench. They found that fresh-appearing basalts are dominated by smectite, sulphides, calcite and talc, secondary minerals associated with nonoxidative (hydrothermal) alteration that probably occurred tens of millions of years ago near the crest of the Galapagos Rise. Many of these nonoxidatively altered rocks are not undergoing oxidative alteration as a consequence of their recent tectonic exposure to ambient seawater. Celadonite and iron oxides are the stable secondary mineral phases.

Origin and Distribution
of Metalliferous Sediment

The distribution and origin of metalliferous deposits of the Nazca plate have been studied extensively by several investigators [Dymond and Ecklund, 1978; Heath and Dymond, 1977; Dymond et al., 1977; Corliss et al., 1976; Dymond et al., 1976; Dymond and Veeh, 1975; Dymond et al., 1973; Dasch et al., 1971] (Figure 3). According to Dymond et al. [1973] sediments from near the basement of a number of Deep Sea Drilling Project (DSDP) sites from the Bauer Basin and from the East Pacific Rise have unusually high transition metal-to-aluminum ratios. Similarities in the chemical, isotopic, and mineralogical compositions of these deposits point to a common origin. All the sediments studied have rare-earth-element (REE) patterns strongly resembling the pattern of sea water, implying either that the REE's were coprecipitated with ferromanganese hydroxyoxides (hydroxyoxides denote a mixture of unspecified hydrated oxides and hydroxides), or that they are incorporated in small concentrations of phosphatic fish debris found in all samples. Oxygen isotopic data indicate that the metalliferous sediments are in isotopic equilibrium with sea water and are composed of varying mixtures of two end-member phases with different oxygen isotopic compositions: an iron-manganese hydroxyoxide and an iron-rich montmorillonite, and various manganese hydroxyoxides are the dominant phases present. Sr^{87}/Sr^{86} ratios for the metalliferous sediments are indistinguishable from the Sr^{87}/Sr^{86} ratio in modern sea water. Since these sediments were formed 30 to 60 m.y. ago, when sea water had a lower Sr^{87}/Sr^{86} value, the strontium in the poorly crystalline hydroxyoxides must be exchanging with interstitial water in open contact with sea water. In contrast, uranium isotopic data indicate that the metalliferous sediments have formed a closed system for this element. The sulfur isotopic compositions suggest that sea water sulfur dominates these

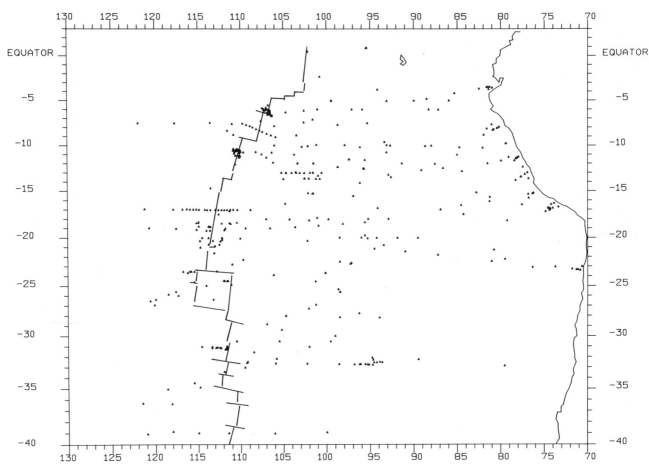

Fig. 3. Location map of sediment samples collected on the Nazca plate. See Figure 1 for location of physiographic features.

sediments with little or no contribution of magmatic or bacteriologically reduced sulfur. In contrast, ratios of lead isotopes in the metalliferous deposits resemble values for oceanic tholeiite basalt, but are quite different from ratios found in authigenic marine manganese nodules. Therefore, lead in the metalliferous sediments appears to be of magmatic origin.

The combined mineralogical, isotopic, and chemical data for these sediments suggest that they formed from hydrothermal solutions generated by the interaction of sea water with newly formed basalt crust at mid-ocean ridges. The crystallization of solid phases took place at low temperatures and was strongly influenced by sea water, which was the source for some of the elements found in the sediments.

Using the chemical data, the origin of metal-rich deposits can be quantified according to contributions from hydrothermal, hydrogenous, biogenic and detrital sources. According to Heath and Dymond (1977), Fe, Mn, Cu, Zn, Ni, Ba, Si and Al are more than 50 percent hydrothermal in East Pacific Rise samples from 10° to 25°S lati-

tude. The first four elements are dominantly hydrothermal in the Bauer Basin and Central Basin as well. Seventy to 80 percent of the Ni, 60 to 80 percent of the Ba, and 30 to 60 percent of the Cu and Zn in the Bauer Basin and Central Basin sediments are hydrogenous. Silicon, Ba, and Zn are dominantly biogenous on the northern East Pacific Rise crest, where more than one-third of the Cu also is derived from this source. Detrital Al and Si are dominant away from the rise crest, particularly in the Yupanqui Basin, where about 40 percent of the Fe and 15 percent of the Zn may also be detrital. Much of the hydrothermal Fe and biogenous Si have been transformed to an iron-rich smectite. The proportion of total Fe bound in this phase varies from less than 20 percent on the southern rise crest to about 40 percent in the Bauer Basin.

The distribution of each element is governed by (1) supply from the four basic sources; (2) lateral transport by bottom currents moving east and then south across the northern East Pacific Rise and Bauer Basin to the Central Basin and moving west from the Peru Basin to the

Central Basin and (3) transformation of the unstable metalliferous hydroxides into more stable smectite and ferromanganese oxyhydroxides.

Crustal Structure, Stratigraphy and Tectonics, of the Andean Forearc and Trench

Peru-Chile Trench

As the Nazca Plate bends and descends into the Peru-Chile trench it undergoes changes in stress which are reflected in structural changes in the uppermost part of the subducting lithosphere. Such structures have been studied extensively [Kulm et al., 1973, Prince and Kulm, 1975, Schweller and Kulm, 1978 and Schweller et al., 1981]. These investigators have described the morphology and shallow structure of the Peru-Chile Trench axis and the down-bending oceanic plate just prior to subduction. Five morpho-tectonic provinces (4°-12°, 12°-17°, 17°-28°, 28°-33°, 33°-45°S) show distinct changes in trench depth, axial sediment thickness, oceanic plate fault structures and dip of the seaward trench slope [Schweller et al., 1981] (Figure 1). The northern and southern regions off northern and central Peru and southern Chile, respectively, are characterized by relatively shallow axial depths, moderate to thick trench axis turbidites, and a gently dipping seaward trench slope with minor subduction-related faulting. In contrast, the deeper central area is almost barren of axial sediments and bends downward more steeply prior to subduction. It also displays an extensive network of major faults with up to 1000 m of vertical offset on the seaward slope.

Two systems of faulting occur in conjunction with subduction. Bending of the oceanic plate causes extensional stress and brittle failure of the upper oceanic crust, resulting in step faults, grabens and tilted fault blocks on the seaward trench slope [Schweller and Kulm, 1978]. Extensional faulting begins near the outer edge of the trench and develops progressively toward the trench axis. However, basaltic ridges and tilted, uplifted axis sediments at several locales along the trench can be better explained by thrust faulting. Compressional stress due to plate convergence occasionally can be transmitted seaward from beneath the continental margin through the oceanic plate, emerging as thrust faulting of the oceanic crust in the trench axis [Prince and Kulm, 1975]. Axial turbidites are commonly tilted landward as they are uplifted, probably as a result of downward curving of the underlying thrust fault [Prince and Schweller, 1978]. Faulting of the oceanic crust prior to and during subduction may have important implications for evolution of convergent continental margins.

The inclined seismic zone is segmented based upon the positions of the hypocenters of earthquakes in the subducting slab beneath Peru and Chile [Barazangi and Isacks, 1976; Stauder,

1975]. While the evidence is not conclusive, the tear zones between adjacent slabs may be manifested by the unusually large vertical offsets (e.g. up to 1 km) in the volcanic basement (layer 2) of the Chile portion of the trench at about 27.5°S and 33°S latitude [Schweller et al., 1981] (Figure 1).

Andean Forearc

The nature of the oceanic crust prior to subduction in the Peru-Chile trench was described in the previous section of this report. On the Peru continental margin layer 2 is traced about 60 km landward beneath the continental slope and displays a broken character in the deep multichannel seismic records obtained at three locations (3.5°S, 9°S, 12°S) off northern and central Peru [Hussong et al., 1976; Kulm et al., 1981; Shepherd, 1979]. In northern Chile a magnetic depth to source study [Blakely, 1981] shows strong attenuation of the magnetic anomalies as the oceanic crust enters the subduction zone and dives beneath the lower continental slope with inferred rupture of the crust underneath the forearc.

Sediment studies of the lower continental slope (i.e. outer forearc) suggest that some oceanic plate deposits are being accreted to the continent, especially where the Nazca Ridge intersects the continent and possibly in other areas off central Peru [Kulm et al., 1974; Rosato, 1974]. On the other hand, the lack of sedimentary basins and the presence of high velocity rocks near the edge of the slope off northern Chile suggest that the oceanic plate deposits and the continental slope desposits are being consumed (tectonically removed) in the subduction process [Kulm, et al., 1977].

Higher on the continental slope and along the outer edge of the continental shelf the subduction complex (i.e. accreted prism) and normal slope deposits interface with the metamorphic crystalline block (i.e. arc massif) that serves as the foundation of the Andean continental block [Jones, 1979; Kulm et al., 1981]. This block is characterized by a high velocity crust (>5.5 km/sec) which is marked by two industry drill holes on the outer edge of the shelf at about 9°S. Crustal structure suggests the block terminates about 10 km seaward of the drill holes where rather extensive occurrences of Late Cenozoic dolomicrite and micrite were dredged along the upper continental slope and outer shelf [Kulm et al., 1981a]. They exhibit measured velocities of 4.5 to 6.6 km/sec which are similar to the drilled metamorphic biotite gneiss and phyllite. These lithologies are absent in the drill holes and along the Peruvian coast. Off Lima, Peru, (12°S) the benthic forams in these dolomites and carbonates indicate subsidence of the upper slope ranging from 500 to 1100 meters [Kulm et al., 1981a]. However, at 9°S and 8°S no subsidence or uplift is indicated by these fauna. Judging from the extensive occurrence of these lithologies, we believe that much of the Lima forearc basin deposits consist of carbonates which suggests that this basin may have a higher potential for petroleum source rocks than those basins containing largely volcaniclastic deposits typical of forearcs.

The crustal structure of the Lima basin and the Arequipa Basin off southern Peru is known largely through seismic reflection studies by Johnson et al. (1975) and Kulm et al. (1981a). Depth sections of the Lima basin shows from 2-4 km of sediments overlying what appears to be the crystalline block and farther seaward the subduction complex. Velocities from multichannel seismic data range from about 1.6 to 4.5 km/sec for the Lima basin deposits (9°S and 12°S) with a 5.5 km/sec material (determined from seismic refraction data) below. Farther south single channel seismic reflection and wide angle reflection data were taken over the trench slope and shelf between 12° and 17°S [Johnson et al., 1975]. Analysis of these data indicate two northeast-trending zones of structural disturbance which intersect the trench at 13.2°S and 16.7°S. The northern boundary zone consists of a series of offsets in structural highs and basins extending from the trench axis toward shore. Upper slope basins, lying at water depths of 1400 m, are prominent north of the 13.2° disturbance and the southern portion of the Lima basin seaward of the Paracas Peninsula appears to be uplifted. Wide angle reflection profiles extend the depth of investigation to 3 km and indicate sediment velocities which increase with depth of burial from 1.5 to 2.5 km/sec. Basement velocities on the shelf range from 4.2 to 5.8 km/sec and range between 5.0 to 5.6 km/sec beneath the upper slope [Jones, 1979].

Johnson et al. (1975) note that the southern boundary zone is less well defined at shallow depths. It extends east-northeast from an abrupt change in character in the trench axis at 16.7°S. The trench axis is broken up and increases in depth very rapidly from 6.8 to 7.4 km. Upper slope basinal structure which is absent over the nose of the Nazca ridge reappears with the presence of a large upper slope basin (Arequipa Basin) at a depth of 1600 m near 17.3°S, 72.5°W. The northern portion of the basin is disrupted, bordered at places by a seaward ridge, broken internally by faults and scoured by subsea channels which run subparallel to the coastlines, perhaps in tectonically-formed troughs in the sedimentary layer.

Between these two zones of disruption lies the Nazca ridge which meets the Peru trench near 15°S [Johnson et al., 1975]. Bathymetrically the ridge is approximately 300 km wide and extends seaward from the trench toward the southwest for more than 1000 km. It rises 2000 m above the normal sea floor on either side of its axis. The topographic expression of the ridge in the trench is less, about 1200 m, and

its presence does not seem to affect trench curvature. The principal effect of the ridge on the margin is to steepen the gradient of the continental slope. In addition, antiformal structures present on the lowermost slope elsewhere are quite apparent near the Nazca ridge. The exceptional thickness of the ridge contributes to a steep subduction angle which in turn results in more scrapeoff of sediments at the trench.

The nature of the deeper crustal structures beneath the Andean forearc are described by gravity models which are for the most part, constrained by seismic refraction data. As noted in the work of Couch and Whitsett (1981), a high density contrast apparently occurs between the Tertiary sediments (2.0-2.5 gm/cm^3 and the older pre-Tertiary units (2.75 gm/cm^3). Consequently, crustal models constrained by gravity, seismic reflection and/or refraction and geologic data yield a good estimate of the volume of sedimentary materials overlying the crystalline basement. These sediments are interpreted as accreted to the continent. Computations based on the rate of plate convergence along the Peru-Chile margin and the thickness of sediments on the oceanic plate indicate that more of the oceanic sediments may be subducted beneath the continental margin than are accreted to it. The percentage of the oceanic sediments subducted apparently varies along the continental margin and is probably indicative of local variations in the subduction process.

Part II: Future Areas of Research

Oceanic Crust

Although oceanographic institutions and the Deep Sea Drilling Program have attempted to sample oceanic crust from the Nazca plate and its boundaries, the results have been less than satisfactory. Large segments of the East Pacific Rise, the Chile Rise and the Galapagos spreading center have not been sampled systematically. Unless this is done in the next decade, detailed examination of along-ridge variations in magma compositions, including those associated with prominent fracture zones, cannot be evaluated. The East Pacific Rise at 20-30°S is the most rapidly spreading portion of the mid-ocean ridge system of the world's oceans, yet we know surprisingly little about the petrogenesis of basalts recovered from this area or the width of the zone of active volcanism. High resolution sampling and observations, either by submersible or by acoustically navigated systems, should be used. Efforts should focus on determining the nature and causes of the compositional variability of basalts recovered from small areas within the region of most rapid spreading. A long-term goal should be the development of a data base and models of crustal formation such as those presently available for slow spreading mid-ocean ridges.

A multidisciplinary approach, involving geophysical experiments (e.g. seismic reflection, refraction, gravity, magnetics), should be used in addition to required geochemical and petrological investigations.

Existing knowledge about the compositional variability of Nazca Plate crust along mantle flow lines or with depth in the section should be improved. Crustal drilling during DSDP Leg 34 penetrated only the upper few tens of meters of layer 2 at three sites. Unless future drilling efforts can extend our sampling to mid and lower layer 2 and upper layer 3, high resolution sampling of natural "windows" into oceanic crust (e.g., fracture zones and fault scarps) should be encouraged. We need to further examine the hypothesis that oceanic crustal sections formed at fast spreading centers reflect derivation from a large steady-state magma reservoir, rather than from small discrete magma reservoirs as has been suggested recently for slow spreading centers. More study should also focus on the nature and degree of alteration of oceanic crust with depth in the section. We understand reasonably the mineral and chemical consequences of oxidative seawater and nonoxidative hydrothermally-induced alteration of surficial layer 2 basement exposures, but the higher temperature alteration of mid and lower crustal rocks remains to be studied.

Nazca Plate Sediments

The metalliferous sediments found on the Nazca Plate have received considerable attention in the past decade. Through isotopic studies and various circumstantial evidence, it seems clear that these deposits, which contain unusually high abundances of iron, manganese, and other transition metals, were formed by precipitation from seawater hydrothermal systems. The studies thus far have been important in determining the present-day distribution of metalliferous sediments on the plate and the influence of other sources on the composition of the metalliferous sediments on the plate and the influence of other sources on the composition of the metalliferous sediments. It is apparent that the Nazca Plate, because of its geographic and tectonic setting, is an excellent laboratory to unravel the chemical and physical processes which control deep-sea sediment compositions. In addition, DSDP Leg 34 permitted an evaluation of the compositional variability and accumulation rates of metals through time and space.

The framework now exists to begin more quantitative studies of the metalliferous deposits as well as the biogenic and detrital-rich sediments in various portions of the plate. With regard to the metalliferous sediments, existing samples were collected during conventional surveying and more detailed sampling in relation to Rise crest tectonic features is required to provide samples that can be used to develop accurate geochemical

budgets. These budgets which could permit an evaluation of the relative rates of hydrothermal input at different parts of the East Pacific Rise, are necessary data for determining the influence of hydrothermal processes for the world's oceans. This type of quantitative data would be important for delineating fruitful areas for more detailed submersible studies. Multibeam bathymetric surveys of large areas of the East Pacific Rise would be an important precursor to these more detailed sampling efforts.

Studies of metalliferous sediments from the East Pacific Rise suggest that the precipitates form from hydrothermal fluids that are chemically distinct from those sampled at the Galapagos Hydrothermal Area. For example, the EPR sediments have a surprisingly constant Fe/Mn ratio of approximately 3.0. The Galapagos Rift fluids have high Mn contents and generally low Fe, presumably reflecting the earlier precipitation of Fe in the form of sulfides. In any case, there appears to be a major difference in the nature of the hydrothermal systems in the two areas. Eventually, sampling and direct observation with submersibles will be necessary to understand the differences in the two areas.

The development of the hydraulic piston core by the Deep Sea Drilling Project has opened the possibility of studying the basal hydrothermal sediments far better than was possible with conventional drilling. Complete core recovery, which is necessary for accurate hydrothermal flux measurements, is difficult for the basal sediments. A suite of hydraulic piston core sites parallel and perpendicular to the EPR would provide valuable data on hydrothermal processes at different times in the past. For example, the shift in spreading from the Galapagos Rise to the EPR should be recorded in the basal metalliferous sediments, and appropriately placed DSDP holes could clarify the timing of this event and any changes in the hydrothermal products.

Andean Forearc

Previous studies have shown that the Andean forearc is extremely complex and that accretion [Coulbourn and Moberly, 1977; Kulm et al., 1977] or consumption [Hussong et al., 1976] of the overiding plate may be occuring at various locations along this forearc. While the metamorphic arc massif is clearly identified in shelf drill holes and on offshore islands, its seaward position beneath the outer edge of the continental shelf or beneath the continental slope is uncertain throughout most of the forearc. For this reason it is difficult to determine how much of the forearc is comprised of the subduction complex which records the history of subduction now being obtained by the JOIDES-IPOD drilling program in other forearcs in the Pacific Ocean. Drilling is the only means of investigating the complete historical record.

Interestingly, the Late Cenozoic sedimentary rocks in the forearc off Lima, Peru, document that subsidence is the dominant motion in this region today. It is similar to the subsidence noted in the drill holes in the Japan forearc [von Huene, Nasu and others, 1978]. However, this portion of the Peruvian forearc exhibits compressional structural features and deformed carbonates which suggest a prior history of compression (probably accretion) before the onset of subsidence in late Miocene to Middle Pleistocene time. Such complicated movements require more detailed geological and geophysical studies of the forearc followed by a comprehensive drilling program.

Future studies should identify the major tectonic end members of the Andean forearc and relate these to the crustal structure of the Andean continental block and other characteristics of the convergence boundary.

References

Barazangi, M. and B.L. Isacks, Spatial distribution of earthquakes and subduction of Nazca Plate beneath South America, Geology, V. 4, 686-692, 1976.

Blakely, R.J., Estimation of depth to magnetic source using maximum entropy power spectra, with application to Peru-Chile Trench, in Kulm, L.D., Dymond, J., Dasch, E.J., and Hussong, D., eds., Nazca plate: crustal formation and Andean convergence, Geol. Soc. Amer. Mem. (in press), 1981.

Bostrom, K. and M.N.A. Petersen, The origin of aluminum-poor ferromanganoan sediments in areas of high heat flow on the Pacific Rise, Mar. Geol., V. 7, 427-447, 1969.

Bostrom, K., Precipitates from hydrothermal exhalations on the East Pacific Rise, Econ. Geol., V. 61, 1258-1265, 1966.

Corliss, J.B., J. Dymond, C. Lopez, Elemental abundance patterns in Leg 34 rocks, in Yeats, R.S., S.R. Hart, et al., eds., Initial Reports of the Deep Sea Drilling Project, Volume 34, pp. 293-300, U.S. Govt. Printing Office, Washington, D.C., 1976.

Couch, R., and R. Whitsett, Structures of the Nazca Ridge and continental shelf and slope of southern Peru, in Kulm, L.D., Dymond, J., Dasch, E.J., and Hussong, D., eds., Nazca plate: crustal formation and Andean convergence, Geol. Soc. Amer. Mem. (in press), 1981.

Coulbourn, W.T. and R. Moberly, Structural evidence of the evolution of forearc basins off South America, Canad. Jour. Earth Sci., V. 14, 102-116, 1977.

Dasch, E.J., J. Dymond, G.R. Heath, Isotopic analysis of metalliferous sediments from the East Pacific Rise, Earth Planet. Sci. Lett., V. 13, 175-180, 1971.

Dymond, J. and W. Ecklund, A microprobe study of metalliferous sediment components, Earth Planet. Sci. Lett., V. 40, 243-251, 1978.

Dymond, J., J.B. Corliss, G.R. Heath, History of metalliferous sedimentation at DSDP Site 319 in the southeastern Pacific, Geochim Cosmochim Acta, V. 41, 741-753, 1977.

Dymond, J., J.B. Corliss, R. Stillinger, Metalliferous sediments from Sites 319, 320 and 321, in Yeats, R.S., S.R. Hart, et al., eds., Initial Reports of the Deep Sea Drilling Project, Volume 34, pp. 575-588, U.S. Govt. Printing Office, Washington, D.C., 1976.

Dymond, J., and H.H. Veeh, Metal accumulation rates in the southeast Pacific and the origin of metalliferous sediments, Earth Planet Sci. Lett., V. 28, 13-22, 1975.

Dymond, J., J.B. Corliss, G.R. Heath, C.W. Field, E.J. Dasch, and H.H. Veeh, Origin of metalliferous sediments from the Pacific Ocean, Geol. Soc. Amer. Bull., V. 84, 3355-3372, 1973.

Ericksen, G.E., Metallogenic provinces of southeaster Pacific region, in Halbouty, M.T., Maher, J.C. and Lian, H.M., eds., Circum-Pacific Energy and Mineral Resources, Amer. Assoc. Petroleum Geologists Memoir 25, pp. 527-538, 1976.

Fisher, R.L. and R.W. Raitt, Topography and structure of the Peru-Chile trench, Deep Sea Res., V. 9, 423-443, 1962.

Getts, T.R., Gravity and tectonics of the Peru-Chile trench and eastern Nazca Plate, 0°-33°S, (Masters thesis), University of Hawaii, Honolulu, 1975.

Handschumacher, D.W., Post-Eocene plate tectonics of the eastern Pacific, in Sutton, G.H., M.H. Manghnani, R. Moberly, eds., The Geophysics of the Pacific Ocean Basin and its Margin, Amer. Geophys. Union Monograph, V. 19, 177-203, 1976.

Hayes, D.E., A geophysical investigation of the Peru-Chile trench, Mar. Geol., V. 4, 309-351, 1966.

Heath, G.R. and J. Dymond, Genesis and diagenesis of metalliferous sediments from the East Pacific Rise, Bauer Deep and Central Basin, Northwest Nazca Plate, Geol. Soc. Amer. Bull., V. 88, 723-733, 1977.

Herron, E.M., Sea-floor spreading and the Cenozoic history of the east-central Pacific, Geol. Soc. Amer. Bull., V. 83, 1671-1692, 1972.

Hussong, D.M., P.B. Edwards, S.H. Johnson, J.F. Campbell, and G.H. Sutton, Crustal structure of the Peru-Chile Trench, 8°-12°S latitude, in Sutton, G.H., M.H. Manghnani, R. Moberly, eds., The Geophysics of the Pacific Ocean Basin and its Margin, Amer. Geophys. Union Monography, V. 19, 71-85, 1976.

Johnson, S.H., G.E. Ness, and K.R. Wrolstad, Shallow structures and seismic velocities of the southern Peru margin, EOS (Trans. Amer. Geophys. Union), V. 56, 443, 1975.

Jones, P.R. III, Seismic ray trace techniques applied to the determination of crustal structures across the Peru continental margin and Nazca Plate at 9°S latitude, (Ph.D. dissertation), Oregon State University, Corvallis, 1979.

Kelleher, J.A., Rupture zones of large South American earthquakes and some predictions, J. Geophys. Res., V. 77, 2087-2103, 1972.

Kulm, L.D., R.A. Prince, W. French, S.H. Johnson, and A. Masias, Crustal structure and tectonics of the central Peru continental margin and trench, in Kulm, L.D., Dymond, J., Dasch, E.J., and Hussong, D., eds., Nazca plate: crustal formation and Andean convergence, Geol. Soc. Amer. Mem. (in press), 1981.

Kulm, L.D., H.J. Schrader, J.M. Resig, T.M. Thornburg, A. Masias, and L. Johnson, Lithostratigraphy, biostratigraphy and paleo-environments of late Cenozoic carbonates on the Peru-Chile continental margin, in Kulm, L.D., Dymond, J., Dasch, E.J., and Hussong, D., eds., Nazca plate: crustal formation and Andean convergence, Geol. Soc. Amer. Mem., (in press), 1981a.

Kulm, L.D., W.J. Schweller, and A. Masias, A preliminary analysis of the subduction processes along the Andean continental margin, 6° to 45°S, in Talwani, M. and W.C. Pitman, eds., Island Arcs, Deep Sea Trenches and Back-Arc Basins, Amer. Geophys. Union, Maurice Ewing Series 1, 285-301, 1977.

Kulm, L.D., J.M. Resig, T.C. Moore, Jr., and V.J. Rosato, Transfer of Nazca Ridge pelagic sediment to the Peru continental margin, Geol. Soc. Amer. Bull., V. 85, 769-780, 1974.

Kulm, L.D., K.F. Scheidegger, R.A. Prince, J. Dymond, T.C. Moore, Jr., and D.M. Hussong, Tholeiitic basalt ridge in the Peru trench, Geology, V. 1, 11-14, 1973.

Mammerickx, J., E. Herron, and L. Dorman, Evidence for two fossil spreading ridges on the southeast Pacific, Geol. Soc. Amer. Bull., V. 91, 263-271, 1980.

Mammerickx, J., R.N. Anderson, H.W. Menard, and S.M. Smith, Morphology and tectonic evolution of the east-central Pacific, Geol. Soc. Amer. Bull., V. 86, 111-118, 1975.

Masias, J.A., Morphology, shallow structure, and evolution of the Peruvian continental margin, 6° to 18°S (Master's thesis), Oregon State University, Corvallis, 1976.

Prince, R.A. and W.J. Schweller, Dates, rates and angles of faulting in the Peru-Chile Trench, Nature, v. 271 743-745, 1978.

Prince, R.A., and L.D. Kulm, Crustal rupture and the initiation of imbricate thrusting in the Peru-Chile trench, Geol. Soc. Amer. Bull., V. 86, 1639-1653, 1975.

Rea, D.K., Tectonics of the Nazca-Pacific divergent plate boundary, in Kulm, L.D., Dymond, J., Dasch, E.J., and Hussong, D., eds., Nazca plate: crustal formation and Andean convergence, Geol. Soc. Amer. Mem., (in press) 1981.

Rea, D.K., Asymmetric sea-floor spreading and a non-transform axis offset: The East Pacific Rise 20°S survey area, Geol. Soc. Am. Bull., V. 89, 836-844, 1978.

Rea, D.K., Evolution of the East Pacific Rise between 3°S and 13°S since the middle Miocene, Geophys. Res. Lett., V.5, 561-564, 1978a.

Rea, D.K., Local axial migration and spreading rate variations, East Pacific Rise, 31°S, Earth Planet. Sci. Lett., V. 34, 78-84, 1977.

Rea, D.K., Analysis of a fast-spreading rise crest: The East Pacific Rise, 9° to 12°South, Mar. Geophys. Res., V. 2, 291-313, 1976.

Rea, D.K., Changes in the axial configuration of the East Pacific Rise near 6°S during the past 2 m.y., Jour. Geophys. Res., V. 81, 1495-1504, 1976a.

Rea, D.K., Model for the formation of the topographic features of the East Pacific Rise crest, Geology, V. 3, 77-80, 1975.

Rea, D.K., and R.J. Blakely, Short-wavelength magnetic anomalies in a region of rapid sea-floor spreading, Nature, V. 255, 126-128, 1975.

Rea, D.K. and B.T. Malfait, Geologic evolution of the northern Nazca Plate, Geology, V. 2, 317-320, 1974.

Rosato, V.J., Peruvian deep-sea sediments: Evidence for continental accretion (Master's thesis), Oregon State University, Corvallis, 1974.

Scheidegger, K.F. and J.B. Corliss, 1981, Petrogenesis and secondary alteration of upper layer 2 basalts of the Nazca plate, in Kulm, L.D., Dymond, J., Dasch, E.J., and Hussong, D.M., eds., Nazca plate: crustal formation and Andean convergence, Geol. Soc. Amer. Mem., (in press), 1981.

Scheidegger, K.F., L.D. Kulm, J.B. Corliss, W.J. Schweller, and R.A. Prince, Fractionation and mantle heterogeneity in basalts from the Peru-Chile trench, Earth Planet. Sci. Lett., V. 37, 409-420, 1978.

Scheidegger, K.F. and D.S. Stakes, Mineralogy, chemistry and crystallization sequence of clay minerals in altered tholeiitic basalts from the Peru trench, Earth Planet. Sci. Lett., V 36, p. 413-422.

Scholl, D.W., R. von Huene, and J.B. Ridlon, Spreading of the ocean floor: Undeformed sediments in the Peru-Chile trench, Science, V. 159, 869-871, 1968.

Scholl, D.W., M.N. Christensen, R. von Huene, and M.S. Marlow, Peru-Chile trench sediments and sea-floor spreading, Geol. Soc. Amer. Bull. V. 81, 1339-1360, 1970.

Schweller, W.J. and L.D. Kulm, Extensional rupture of oceanic crust in the Chile trench, Mar. Geol., V 28, 271-291, 1978.

Schweller, W.J., L.D. Kulm, R.A. Prince, Tectonics, structure and sedimentary framework of the Peru-Chile Trench, in Kulm, L.D., Dymond, J., Dasch, E.J., Hussong, D.M., eds., Nazca plate: crustal formation and Andean convergence, Geol. Soc. Amer. Mem., (in press), 1981.

Shepherd, G.A., Shallow structure and marine geology of a convergence zone, northwest Peru and southwest Ecuador (Ph.D. dissertation), Oregon State University, Corvallis, 1979.

Shor, G.G. Jr., H.W. Menard and R.W. Raitt, Structure of the Pacific basin, in Maxwell, A.E., ed. The Sea, V. 4, New York, Wiley Interscience, 3-28, 1970.

Stauder, W., Subduction of the Nazca plate under Peru as evidenced by focal mechanisms and by seismicity, J. Geophys. Res., V. 80, 1053-1064, 1975.

von Huene, R., N. Nasu, et al., Japan trench transected, Geotimes, V. 23, 16-20, 1978.

Whitsett, R.M., Gravity measurements and their structural implications for the continental margin of southern Peru (Ph.D. dissertation), Oregon State University, Corvallis, 1976.

Worzel, J.L., Standard oceanic and continental structure, in Burk, C.A. and Drake, C.L., eds., The Geology of Continental Margins, New York, Springer-Verlag, V. 1, 8, 1974.

Yeats, R.S., S.R. Hart, et al., Initial Reports of the Deep Sea Drilling Project, V. 34, Washington, D.C., U.S. Govt. Printing Office, 814, 1976.

Zeigler, J.M., W.D. Athearn, and H. Small, Profiles across the Peru-Chile Trench, Deep-Sea Res., V. 4, 238-249, 1957.

GEOPHYSICAL DATA AND THE NAZCA–SOUTH AMERICAN SUBDUCTION
ZONE KINEMATICS: PERU–NORTH CHILE SEGMENT

L. Ocola

Applied Geophysics Branch, Instituto Geofísico del Perú, Apartado 3747, Lima-PERU

Abstract.   Results from deep seismic sounding show that the continent-ocean transition zone structure is highly complex along the western boundary of the South American Plate. The main results are: i) 5.0-5.4, 6.0-6.3, 7.0-7.3, and 7.9-8.3 km/sec refractors under the deep ocean, ii) the average dip of the crust-upper mantle transition is about 6° towards the continent, iii) on the continental shelf off-central Peru the refractor velocity families are: 5.1-5.3, 6.0-6.2, 6.8, 7.3 km/sec for the material below the sediment cover.   Off-southern Peru, models satisfying travel times and wave forms show an upper crust with velocity increasing with depth from about 6.0 to 6.3 km/sec, a lower crust with velocity about 6.9 km/sec and a velocity inversion immediately above the Mohorovicic discontinuity.

A regional gravitational anomaly map, reduced to sea level datum, shows great amplitude anomalies associated with regional tectonic, geologic and morphologic elements. The Nazca ridge, on the other hand, has no gravitational signature, i.e., it is isostatically compensated, but the electrical conductivity data shows a significant anomaly on its extension under the continent.

From the integration of seismicity, geochronology, geochemical, geologic and tectonic data, a model for the Andean subduction zone is proposed. This model considers the coexistence of two normal dipping Benioff zones, i.e., dipping 30° for central Peru, and one normal Benioff zone for southern Peru-north Chile. The western zone is a young relatively shallow belt, presently developing, and it is associated with the Peru trench in central Peru. The eastern zone is an old relatively deep Benioff zone, and it is in process of extinction. This zone is, approximately, along the Subandean zone. Both zones are connected through a low dip, seismically active detachment zone. The episodic nature of intrusive implacements is related to periods of re-activation of South Mid Atlantic ridge and to the vertical migration of magmatic blobs into the South American Plate upper-mantle and crust. The final stages of this emplacement are assumed to be guided by pre-existing structures, so that great bulges on the earth surface will not be produced. Taking into account the geochronological data for the Pucallpa alkalic complex. the approximate date of the third active spreading episode of the South Mid Atlantic spreading center, and the average westwards velocity of displacement of the South American Plate, 4.4 cm/yr ascent velocity is calculated for a 5-km radius spherical blob moving up in a medium with 5×10^{20} poise of viscosity for 300 km path.

Introduction

Knowledge of subduction zone kinematics is of most importance in understanding Earth's dynamics and causal processes. A great deal of effort by earth scientists has been devoted to this end. At present, a wealth of data have been gathered and interpreted, as a result, a great number of models has been proposed to explain observed facts or inferred material properties or processes taking place deep on earth.

Processes of prime importance on the earth's history are those that take place in zones of plate convergence, in particular, in zones like the Andean one which was thought to be a typical paradigm, in the early years of plate tectonic theory. However, as more and better data is collected, proposed models differ from the typical ones.

In the first part of this paper, geophysical data on material properties, and dynamics of the Nazca-South American plate

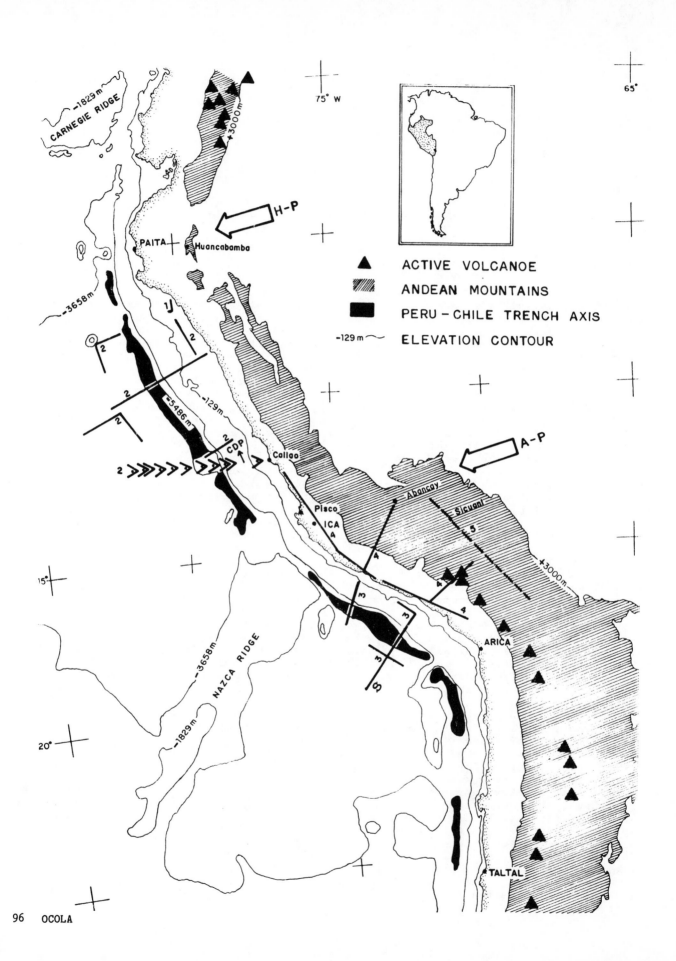

ACTIVE VOLCANOE

ANDEAN MOUNTAINS

PERU – CHILE TRENCH AXIS

−129 m ⁓ ELEVATION CONTOUR

subduction zone: Peru segment, for the
period 1973-1979, are briefly reviewed. A
similar review can be found in Woollard
and Ocola (1973), for years before 1973.
In the second one, models that explain
geologic and tectonic gross features, and
seismicity are proposed. The models take
also into account, geochemical data,   the
westward displacement of South American
plate, the great linear features of
regional tectonic and geologic elements ;
and attempt to explain present seismicity
morphology, volcanic activity distribution
and other tectonic manifestations such as
the great elevation and bulk of the south
Peru Andes (including the Peru-Bolivia
altiplano), marine terraces, etc.

Deep Seismic Sounding

During the last decade, the controlled
source seismology community has maintained
a great interest in determining   material
properties and structure of the crust and
upper mantle in the ocean-continent
transition zone   and neighbouring areas
along the western margin of the South
American plate.
In 1976, a multinational seismic
refraction experiment was carried out with
participation of U.S.A.   universities of

TABLE 1.   1976 Multinational Controlled -
Source Seismic Experiment   Shot
Point Location.

Place	Latitude °S	Longitude °W	Comments
Atico	16.42	73.70	Shot placed in the ocean.
Cuajone	17.04	70.70	Open pit copper mine.
Marcona	15.19	75.13	Open pit iron mine.
Toquepala	17.24	70.61	Open pit copper mine.

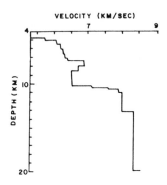

Fig. 2. Velocity-depth function for deep
ocean west of the Peru trench axis which
matches travel times and wave forms
derived by the construction of synthetic
seismograms assuming a plane layer medium.
The velocity for the low velocity zone was
set to 6.5 km/sec (after Meeder et al.,
1977).

Wisconsin, Texas and Washington, Carnegie
Institution of Washington,   and Instituto
Geofísico del Perú.   The lay out of the
experiment is shown in Fig. 1,   lines 3s
and 4s.
In the first part of the experiment
seismic data for several profiles were
gathered between Lima (12°S latitude) and
the Peru-Chile border from four shot
points: three on land, one in the ocean.
The location of these points are given in
Table 1.   The analysis and interpretation
of results have not been published yet.
In the second part, two on-shore and
off-shore profiles were obtained. The land
stations were set along profiles normal to
the Andes. Besides these two lines,
several lines parallel to the trench axis
were fired and data were recorded on ocean
bottom seismometers (OBS).
Meeder et al. (1977) discussed results
from profiles fired on the deep-ocean side
of the trench. They derived several
models using different methods of
interpretation of seismograms assembled in
'record sections'. In general, the models

Fig. 1. Location of controlled source seismic studies reviewed in this paper,   and
major tectonic and morphologic elements in western South American subduction   zone.
The seismic lines are labelled according to (1976) profiles: Solid lines are
standard modalities of seismic refraction profiles. > simbols due west   off-Callao
indicate ASPER stations. CDP: Common-depth-point digital seismic line; 3)   Oceanic
profiles of 1976 multinational experiment (Meeder et al., 1977).   Crustal models
described in this paper correspond to the southernmost line (S); 4) Continental
profiles along- and across-strike of the Andes mountains from shots on land and   in
the ocean (Aldrich et al., 1976), Table 1; 5) Peru-Bolivia altiplano seismic
refraction profile (Ocola et al., 1971). A-P: Abancay-Pisco deflection,   H-P :
Huancabamba-Paita deflection. Note that the active volcanoes (solid triangles) are
outside of the segment bounded by the two deflections.

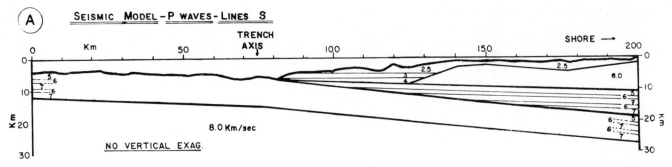

Fig. 3A. Seismic model of Peru trench subduction zone between the shelf and OBS stations. Outlined by heavier lines are oceanic crust and upper mantle, sediment wedge of the continental slope along with sediment basin on the continental shelf, higher velocity material underlying the continental shelf, and a triangular region with crustal crystaline rock velocities. This last region is shown schematically as a thrust of oceanic crust, but could also be continental material (after Keller et al., 1978).

show a monotonically increasing velocity depth function with a low velocity zone in the lower crust. A model, using theoretical travel time computation and synthetic seismogram that closely matched the observed travel time and wave forms of 50-km profile normal to the trench axis is shown in Fig. 2. The main features of this model are the velocity inversions between 8- and 10-km depth, the 7.9 km/sec velocity at Moho discontinuity.

Keller et al. (1978) modeled the deep structure of the shelf and continental slope, southernmost line (S), Fig. 3A. The model was constrained to satisfy reversed travel times between the shelf and deep ocean OBS, near vertical reflection times, and gravity anomaly data. The important features of the model are the thin low velocity layers in two of the deepest three refractors under the continental shelf, the 'low velocity material of the continental slope', and dip about 6°

towards the continent, of the oceanic Moho under the continental shelf.

In 1976, Hussong et al. published a composite crustal section for the ocean-continent transition zone between 8° and 12°S latitude, lines marked with 2 in Fig. 1. According to the Hussong et al. (1976), data from five standard two-ship reversed refraction lines, eleven overlapping split profiles along a 375-km line extending from the shelf to deep ocean across the trench, twelve airgun sonobuoy precision recorder (ASPER) stations distributed on line, were used. The deep crustal structure was obtained from a 190 km long reverse profile parallel to the coastal line, and shot on the flat part of the continental shelf between 8° and 10°S latitude. The refraction data was suplemented by a commercial multichannel, common-depth point (CDP) digital reflection seismic line on the continental slope.

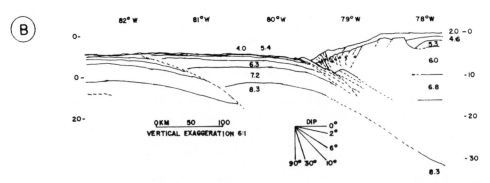

Fig. 3B. Composite crustal cross-section of Peru trench including the major tectonic features identified from seismic data between 8° and 12°S latitude (after Hussong et al., 1976). Note the underthrust fault at about 250-km west from the Peru trench axis.

TABLE 2. Compressional Wave Velocity
Depth Section in the Off-Shore
Part of Sechura Basin (6.9°S,
80.4°W) (Shepherd, 1979).

Depth* km	Velocity km/sec	Geology
0.00	1.5	Water
0.14	1.65	Quaternary
0.51	2.03	U. Tertiary
0.92	2.48	L. Tertiary
1.80	5.08	Mesozoic
5.14	6.22	Paleozoic

(*) depth referred to the upper interface

The traditional pick-and-plot, unreduced travel time method of interpretation was applied to derive crustal models. Fig. 3B is Hussong et al. (1976) generalized composite structural section of Peru trench region and deep ocean derived from "integrating" all crustal models between 8° and 12°S latitude. This section shows oceanic crustal underthrusting on the Nazca plate. Whether such structure is seismically active or not is a question to be answered in the future. For the shallowness of the structure and the major role that water may play on crustal deformation, seismic events associated with this structure might be a small magnitude. Considering the remoteness and the low sensitivity of the closest teleseismic stations no reliable data is available to search for an answer. However, it is without question that compressive stress environments do exist in the oceanic plate. As stated later in this paper, it is proposed that structures of this kind are related to the birth of new Benioff zones.

Crustal model of Meeder et al. (1977), and Hussong et al. (1976) are dissimilar. However, it is surprising that in both models, the 8.3-km/sec refractor begins at about the same depth. Because of the importance of the low velocity layer at the bottom of the crust, it would be interesting to apply Meeder's interpretation procedure to Hussong's et al. data and to investigate evidence for such layer. Keller et al. (1978) model for southern Peru ocean-continent transition should be considered as preliminary. The number of velocity inversions is high. The interpretation needs other constraints besides gravity and reversed travel times, like wave form analysis, for example.

North of Hussong et al. profile, Shepherd (1979) reports Wipperman's results for a 12-km sonobuoy profile shot at about 6.90°S latitude and 80.4°W longitude on the wider part of the continental shelf in the "off-shore part of the N-S trending Sechura basin", Fig. 1-number 1. The results are presented in Table 1. The correlation with Hussong's et al. refractor under the shelf, is in general satisfactory.

Gravity

A regional gravity map of Peru and neighboring ocean covering the active ocean-continent transition zone is presented in Fig. 4. The gravitational anomalies are referred to sea level, thus, the anomalies are the normal Free Air in the ocean, and simple Bouguer on the continent (i.e. Free Air and Bouguer plate corrections applied to the observed gravity). Although the oceanic data from the Nazca Plate Project provides much more information of the fine structure, the regional features remain. Figs. 5 and 6, from Shepherd (1979) and Whitsett (1975) respectively, illustrate this point.

The regional gravitational anomalies cover the country at its full length. From west to east, the main anomalies are:

1) Open ocean: average about zero milligals.
2) Peru-Chile trench with extreme values between -200 mgal, in the southwestern tip of the Nazca ridge, and -150 mgal, in the northwest tip of the Nazca ridge. The anomaly amplitude on the landward extension of the Nazca ridge diminishes significantly on the axis of the trench.
3) Near the coastal line, along the coastal cordillera, a narrow positive anomaly develops with values about +50 mgal in southern Peru, and over +90 mgal in northwest Peru.
4) The Andean anomaly is a dominant feature on the continent with extreme value of about -400 mgal approximately along the axis of the Andean mountains. The axis of the anomaly shows about 200-250 km horizontal offset along the Abancay-Pisco deflection, Fig. 1, with a significant effect on the gravitational anomaly on the axis of the trench off-Pisco, where the trench anomaly becomes more positive, Fig. 6.

In northern Peru, the Paita-Huancabamba deflection have no major effect on the trench axis, Fig. 5. The gravity east of the Andes gradually tends towards the zero value anomaly.

Whitsett's map, Fig. 6, shows the isostatic compensation nature of the Nazca ridge. This effect is absent in ridges off Paita-Huancabamba deflection, where

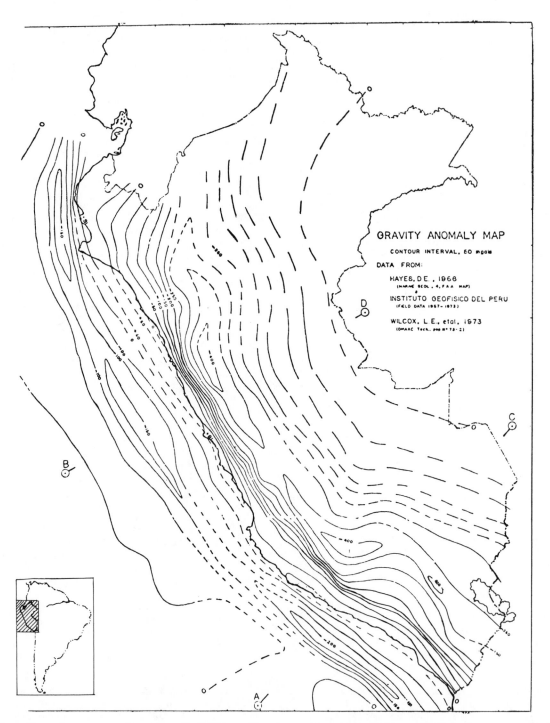

Fig. 4. Gravity anomaly map of Peru. The gravity data was reduced to a sea level datum, with the simple Bouguer plate correction included for the land data, i.e., the anomalies are simple Bouguer on the continent and free air in the ocean.

Fig. 5. Detailed gravitational anomaly (free air) map off-shore northwest Peru, 2.5°– 7°S latitudes. The anomalies produced by deep ocean ridges and shelf capes are strong (after Shepherd, 1979).

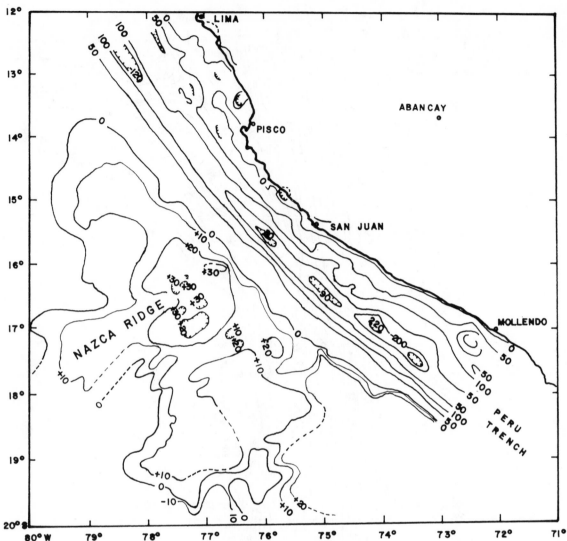

Fig. 6. Detailed gravity anomaly (free air) map off-shore southern Peru. Note the lack of gravimetric expression of the Nazca ridge both under the deep ocean and on the trench axis, i.e., the ridge is isostatically compensated (after Whitsett, 1975).

the gravimetric signature of the topographic elements are striking, Fig. 5.

Electical Conductivity

For more than a decade the Department of Terrestrial Magnetism of Carnegie Institution of Washington, and its South American collaborating institutions gathered data from geomagnetic observations along the Andean mountains and neighboring coastal areas. In 1972, results of the analysis and an interpretation was published as maps showing the co-transfer function between N-S (H) and vertical (Z) components of variations in the geomagnetic field (Aldrich et al., 1972). Their

data have been recontoured, and results are presented in Fig. 7. In the same maps, the 3658- and 4023-m bathimetric contours for the Nazca ridge are shown.

According to Aldrich et al. (loc. cit.), the zero contour line marks, Fig. 7A and 7B, the position of the anomalous material closest to surface, assuming a two dimensional structure. The contour pattern of the 64-min period co-transfer function between the horizontal and vertical components of the geomagnetic field, Fig. 7A, is simple. The anomalies tend to be positive towards the Brasilian shield, and negative towards the Peru trench. However, the H-Z co-transfer function at a period of 16 min, Fig. 7B,

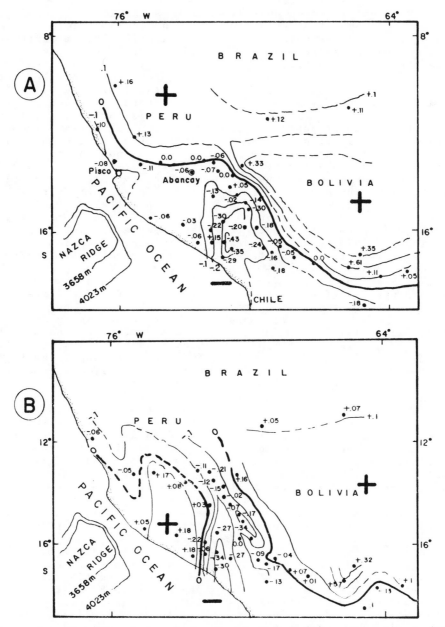

Fig. 7. Co-transfer function between horizontal N-S (H) and vertical (Z) components of variation in the geomagnetic field, data from Aldrich et al. (1972). A. 64-min period, 0.1 units contour interval. B. 16-min period, 0.1 units contour interval. The bathimetric contours of the Nazca ridge are in meters.

has much more structure. There is a well developed positive anomaly in the land extension of the Nazca ridge. This anomaly provides one of first possible evidences for the extension of the Nazca ridge under the continental plate. It would be interesting to map this anomaly in more detail and to complete the study with the appropriate inversion of the data.

As pointed out by Casaverde et al.

(1968), the E-W deflection of the zero anomaly contour line (axis of the anomalous body) is coincident with the Abancay-Pisco deflection (Fig. 1). Hence, both gravitational and geomagnetic data correlate well with the tectonic pattern in this region. The Abancay-Pisco deflection is a major relatively deep tectonic feature; possibly associated with the effect of the continuation of the Nazca

ridge under the continental plate, and it
seems to have played an important role in
the definition of the present Benioff-
subduction-zone morphology and its
evolution on both sides of the deflection
axis.

Seismicity and Tectonics

Several models have been proposed to
explain the spatial distribution of seismicity north of the Nazca ridge (in the
ocean) and the Abancay-Pisco deflection on
the continent and the anomalies of the
volcanic activity along the Andes.

The arguments for or against any
particular model are mainly centred on the
interpretation of the seismicity pattern
(Barazangi and Isacks, 1976, 1979; Stauder,
1975), and models derived from the interpretation of converted and reflected-
refracted seismic phases on what it is
believed to be the upper surface of the
down going slab, on the Q-structure of the
subduction zone (Snoke et al., 1977; Snoke
and Sacks, 1975; Okada, 1974).

In 1979 Barazangi and Isacks selected a
good quality data set from the International Seismological Center (ISC)
catalogue in order to resolve the geometry
of the descending Nazca plate beneath the
central and north Peru. The plan view of
this data is presented, in a modified
format, in Figs. 8-I and 8-II. Fig. 8-I
shows events with depths between 0 and
70 km, and Fig. 8-II displays events with
depths from 70^+ to 320 km. Barazangi and
Isacks (1976) prepared similar plots with
a larger set and lesser quality of data.
The plan view of such data is reproduced
in Figs. 9-I and 9-II in the same format
as Figs. 8.

North of the Arica bight, the 0-70 km
depth seismic activity, Figs. 8-I and 9-I
is distributed in two clear belts: one
along the coast and the other along,
primarily, the Subandean zone. These belts
are the Ica (Nazca)-Paita and the Subandean (Ocola, 1978). The envelopes
around the seismic activity of such belts
are shown in each figure pair.

The following is evident from figures 8
and 9:

1) The 0-70 km depth seismic belts are
not parallel. They seem to converge near
the south east edge of the Nazca ridge.
Hence, the method presenting seismicity
sections by projecting the seismic
activity into a vertical plane
perpendicular to any of the two belts will
give an erroneous picture when block
dimensions parallel to trench axis are
large.

2) The 0-70 km depth coastal belt
follows the general trend of the Peru
trench, the Subandean belt does not.

3) The 70^+-350 km depth seismic activity
shows a definitive segmentation, Figs. 8-II
and 9-II. There are 4-segments. From
south to north these segments are:
i) Taltal (Chile, about 25°S latitude)-
Nazca (Peru near 15°S), segment 4. This
segment is the best developed and its
geometry, taking into account the curvature of the Arica bight, fits the plate
subducted under an island arc paradigm.
Its general dip is about 30°. In this
segment, there are presently active volcanoes in a single active belt approximately parallel to the Peru-Chile trench,
Fig. 1. ii) Nazca transition segment: it
developes along what might be the
continental extension of the Nazca ridge.
It marks the transition between segments
(2) and (1) from segment (4) (Figs. 8-II
and 9-II). This segment includes Barazangi and Isacks tear (Barazangi and Isacks,
1979). However, the morphology of the
transition zone seems to be more gradual
than expected, as shown by Hasegawa and
Sacks (1979). iii) The Subandean segment
(2): this segment is nearly parallel to
the incipiently developed belt along the
coastal area. The activity is not evenly
distributed along the segment. Its dip is
about 30°, and its axis does not parallel
the 0-70 km depth seismic activity belt.
iv) Finally, the fourth segment incipiently
developed, is the Nazca (Ica)-Paita. The
longitudinal axis of this segment, is
approximately parallel to the shallow
activity segment, and follows the general
strike of the Peru trench.

4) There is no significant seismic
activity in the block surrounded by belts 1,
2, and 3.

The spatial distribution of the seismic
activity and the "uniformity" of the geometrical properties of the Taltal-Nazca
segment have been studied in detail by
Barazangi and Isacks (1976, 1979). Among
the important points made by them were
that the thickness of this zone is about
30 km thick, the coherence between the
seismicity in the transverse cross-
sections was remarkable. There is no
major disagreement on this interpretation
(see for example besides Barazangi and
Isacks' paper; Snoke et al., 1977; Snoke
and Sacks, 1975; Rodríguez et al., 1976;
Ocola, 1978).

Two Benioff Zone Model

As shown above, the seismic activity
north of the Abancay-Pisco deflection and
Nazca ridge occurs in multiple belts which
are not all parallel. Hence composite

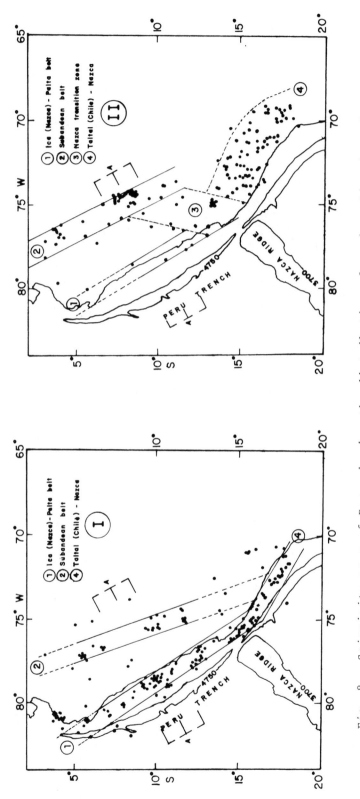

Fig. 8. Seismicity maps of Peru showing the distribution of good-quality data for the period 1964-76 from Barazangi and Isacks (1979, Fig. 3). The envelopes to the seismic activity are the same as Figs. 9-I and II. 8-I: Seismic events with depths between 0 and 70 km. 8-II: Seismic events with depths between 70 and 320 km. Corridor A-A is the same as Barazangi and Isacks' (1979) I-I seismic cross section.

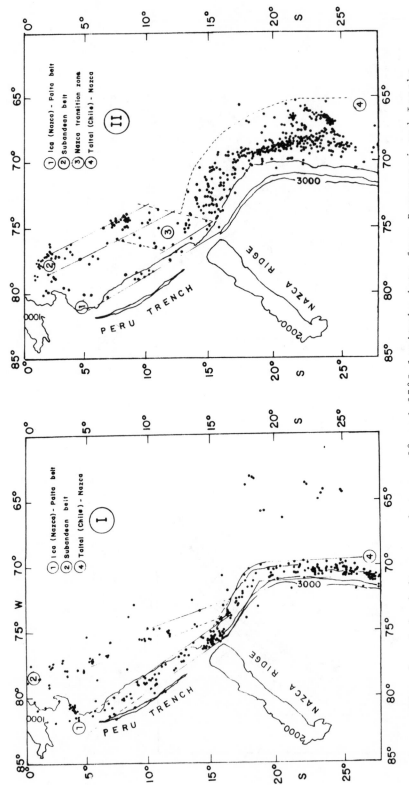

Fig. 9. Seismicity maps between 0° and 27°S latitude data from Barazangi and Isacks (1976, Fig. 1, ISC data). 9-I: Seismic events with depths shallower than 70-km. The seismic belts between the landward projection of the Nazca and Carnegie ridges are not parallel. The three belts: Taltal-Nazca (4), Subandean (2), and Ica (Nazca)- Paita (1) show no obvious segmentation. 9-II: Seismic events with depths between 70 and 320 km. The Subandean seismic activity belt (2) and the Ica-Paita (1) belt tend to be parallel. The Taltal-Nazca belt (4) is very active and well developed. There is a transition belt (3) in the landward extension of the Nazca ridge. The segmentation of the intermediate depth seismic environment is evident.

seismicity cross-sections can not be used to interpret the seismic activity distribution in a vertical plane normal to the general trend of any particular belt or segment, or any tectonic element. The model proposed in this paper, for the seismic activity distribution in the northern segment of Peru is shown in Fig. 10. In this figure, Barazangi and Isacks (1979, Box 1-1, Fig. 3 and Fig. 4: Section 1 class A and B) data is presented together with our preferred interpretation.

In the construction of this model, results from controlled source seismology, and the general dip of 30° have been used for the Benioff slab. The kinematic evolution as well as plausible petrological-geochemical implications are discussed later in this paper. The model considers two Benioff zones. The western-most zone is associated with the present Peru trench. This zone is young and the seismogenic portions of the subducted slab have not reached a great depth yet. The easternmost Benioff zone is an older one and extends primarily under the eastern flank of the Andes. This zone is believed to be in a process of being extinct. Both Benioff zones are related through a low angle "detachment zone", i.e., the "plate" does not continue from the western Benioff zone to the eastern one. The detachment zone becomes shorter in horizontal extension (along the cross-section), as the Arica bight is approached from the north. The importance of this zone in the kinematic evolution of South American will be discussed later. The Barazangi and Isacks 1976-1979 tear, and Hasegawa and Sacks (1979) near horizontal portion of the "plate" are part of the detachment zone.

Crust- Upper Mantle Dynamics Model

Reconstruction models of past western South American tectonism are primarily based on seismological evidence from natural events, and near surface geological information. The models have not fully considered the South American plate mobility, petrological, and geochemical data of rocks which require magmatic material arising from great depth environments, the episodic nature of the tectonic activity, and other geophysical data. In this section, an attempt is made to integrate multidisciplinary data, and interpreted the South American - Nazca plates kinematics in the Andean subduction zone.

The proposed model assumes that the kinematic evolution of the South American-Nazca plates system is a succession of a locked and unlocked stages. The locked stage is when the mobility of South American plate is minimun or insignificant. The unlocked stage, on the other hand, is that for which the South American plate velocity of displacement is significant or reaches a maximum. The locked phase is characterized by a tectonism primarily deformation (compressional). During this phase the coastal uplift is important the bending of the Benioff slab (i.e., maximum curvature) is near the tectonis trench axis, multiple and well developed Benioff zones coexist. The unlocked phase is characterized by a tectonism associated with significant horizontal displacement of the overriding South American plate. This displacement takes place along a deep detachment zone involving the upper-mantle and crust. There is no relationship

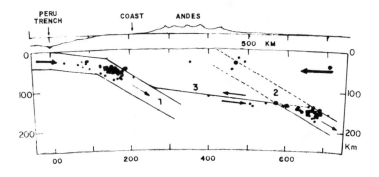

Fig. 10. Seismic cross-section of block A-A, Figs. 9. Data from Barazangi and Isacks (1979, Fig. 4: section1, class A and B): (1) young Benioff zone in process of development, (2) old Benioff zone in process of extinction, (3) detachment zone axis. The arrows indicate direction of displacement.

between the position of the surface contact between the Nazca and South American plates (pseudo trench) and the position of the Benioff zone.

The South American plate displacement speed is controlled primarily by the morphology of the detachment zone, in particular, by its predominant dip (the speed will be greater the smaller the dip), the partial melt zones associated with stress concentrations along the subducted slab, and the rate of subsidance of the oceanic plate (Nazca Plate).

The end of the locked phase is marked by the relatively sudden increase of the over-riding South American Plate displacement speed and the birth of a new Benioff zone under the ocean. The end of the unlocked phase, on the other hand, will begin when the South American Plate meets the sub-ducted slab of the new Benioff zone.

Volcanic activity will developed only where the subducted material reaches the appropriate pressure-temperature, and geo-chemical environments to produce magmatic material.

Kinematics of the Last 14 my

Applying the proposed general model to corridor A-A, Figs. 8-I and 8-II, the last 14 my (my=million years) kinematics of the South American - Nazca plates system is reconstructed. An initial, two intermediate and, one final stage are sketched in Figs. 12 I-IV. It is assumed an average horizontal velocity of displacement 2.2 cm/y for the South American plate, and that the deep section of the subducted slab did not change significantly of position with time. This assumption is justified considering the great thermal and mechanical inertia of the crust and upper mantle directly involved in the sub-duction process. As a consequence of this assumption, the oceanic surface and South American plate contact (Peru trench) was about 308-km east of the present trench position 14 my ago. At this time, the Benioff zone was fully developed, and partial melt zones might have existed in appropriate magmatic environment along the subducted slab, Fig. 12-I.

According to LePichon (1968), 10 my ago, in late Miocene time, the third cycle of spreading of South Atlantic started. This implies that the South American plate was forced to move westwards at a faster rate producing an increase in stress concen-tration along the subducted slab, and accelerating the rate of magma formation. After coalescing the small portions of the melt, the magma formed blobs that burnt their way straight up to the upper crust.

The ascent of the blob into the upper mantle and crust can be numerically modeled by a fluid sphere moving vertically upwards in a high viscosity medium. This problem was solved by Lamb (1832), who found that the ascent velocity (U) was given by:

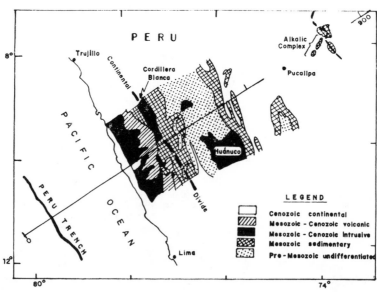

Fig. 11. Regional geologic strip map across the Andes of central Peru (after James, 1979, Fig. 70, modified). This strip coincides approximately with the box A-A of Figs. 8-I and 8-II. The line normal to the trench axis nearly corresponds to seis-micity cross-section axis, Fig. 10.

$$U = \frac{1}{3} \quad a^2 g \Delta\rho \quad \left| \frac{1 + \eta_2/\eta_1}{1 + \frac{3}{2}\eta_2/\eta_1} \right| \qquad (1)$$

where η_1 and η_2 are the viscosity of the imbedding medium (upper-mantle and crust), and sphere (magma), respectively; a: the radius of the sphere, g: gravitational acceleration; and $\Delta\rho$ is the density contrast.

Thus for $\eta_1 >> \eta_2$ equation (1) becomes:

$$U = \frac{1}{3} \quad \frac{a^2 g \Delta\rho}{\eta_1} \qquad (2)$$

According to James (1977), the magma source for the Pucallpa alkalic complex, Fig. 11 should have, at least, been at 300--km depth. Hence, if it assumed that the magma blobs started their ascent inmediately after the South Mid Atlantic ridge reactivation, then the average ascent velocity of the magma blob can be estimated. For the horizontal distance between present implacement of the Pucallpa alkalic complex and the intersection of the 300-km isobath with the subducted slab is about 150-km, this implies that it had taken about 6.82 my to the South American plate to move over this distance. Thus the blobs ascent velocity is about 4.4 cm/y, which is not unreasonable. Furthermore, if it is assumed that the average blob radius is like a Cordillera Blanca's plutons (Pitcher, 1978), Fig. 11; and a viscosity of 5 x 10^{20} poises for the upper-mantle subsolidus, the 4.4 cm/y ascent velocity implies that the density contrast between the magmatic blob and the mantle-and-crust was 0.86 gm/cc, on the average. It is expected, however, that much larger density contrast might have existed when the blob was still in the mantle. The final implacement of the blobs in the upper crust, though can strongly be guided by pre-existing geologic structures such as regional deep faults or zones of weakness.

When the South Mid Atlantic ridge reactivated, underthrust faults might have developed in the Nazca Plate giving birth to the present Benioff zone, Figs.12 I-III. As the time progressed, the subduction process continued along the old Benioff zone, and the South American Plate migrated westward. At depth, stress concentrated on the slab until the material reached a soft shear state and a detachment zone formed. This detachment zone made the westward displacement of the South American Plate easier.

Sections or portions South American Plate segment between Arica bight and immediately north of the Huancabamba-Paita deflection moved faster or for a longer time, the farther these sections were from the bight. The imprint on the geomorphology of the Andes can be appreciated in Fig. 1. The 3000-m elevation above sea level contour shows this dramatic difference. South of the Abancay-Pisco deflection, the locked stage seems on the great extent and altitude of the Peru-Bolivia altiplano and its complicated tectonics and geology. On the other hand, the segment between the deflections does not show similar feature. It appears that the unlocked stage was dominant.

Comparison With Other Models

There are two independent groups of models for the subduction of the Nazca Plate beneath central Peru: The flat geometry model (Barazangi and Isacks,1976), and the normal dipping one (Snoke et al., 1977). In the first model, the Benioff zone dips about 10°, and it is separated from the shallow activity of the upper 50-km of the overriding South American plate. This model does not explain the normal dip (30° towards the continent) of the seismic activity shallower than 100-km close to the trench nor the significant dip of the activity between 100- and 200-km of depth under the Subandean zone. In the second model, the Benioff zone dips 30°and reaches depths greater than 200-km. This model does not explain the activity under and east of the Andes. Hasegawa and Sacks (1979) proposed a conciliatory model by considering the subducted slab deformed into three segments. The first segment includes the seismicity of the upper 100-km and near the trench; the second segment the flat section of the seismic activity under the Andes and, the third one the activity under east of Andes. The dip of the third segment is about 30° eastward. In this model, it is difficult to explain the mechanism thereby the subducted slab attained the double bending. The model here proposed is simpler. It satisfies the seismicity data as well as the conciliatory model does, but takes into account better the mobility of the South American Plate, the episodic emplacement of linear intrusives along the Andean belt, the systematic variation of the length of the seismicity 'flat segment' between the normal dipping shallow activity near the trench and the intermediate depth seismicity under the Subandean region.

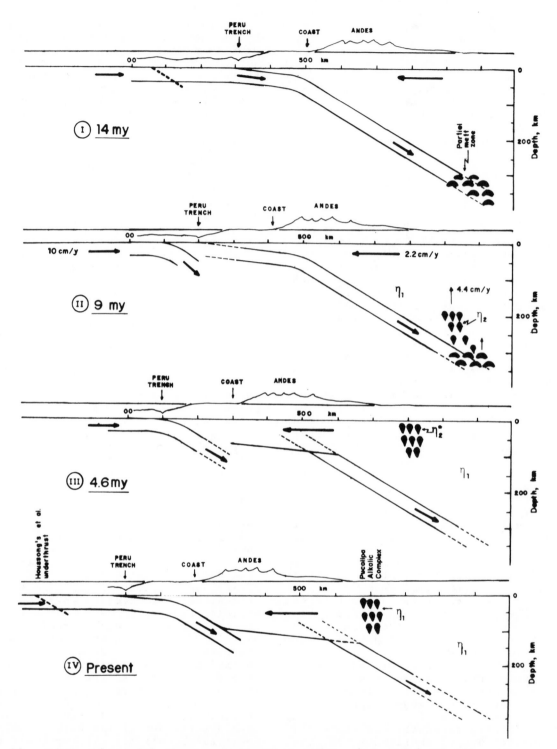

Fig. 12. Kinematic model along corridor A-A, Figs. 8-I and 8-II, diagrams are schematic. I.- Plausible geometry of the Benioff zone 14 my ago. The partial melt zone is one of several possible melting zones along the subducted slab. The heavy solid discontinuous slanted line is potential Hussong's et al. (loc. cit.) under-thrust fault. II.- The reactivation of the South Mid Atlantic ridge in its third (late Miocene) cycle at about 10 my increased the partial melt rate of the subducted material, and blobs of low viscosity material were formed. The blobs burnt their

way up into the upper mantle reaching ultimately near surface. For the continental plate was moving westwards the blobs stream path would look like deflected towards the west. This figure cartoons, besides, the old Benioff zone geometry, the westward displacement of the continental plate, the migration of the "trench", and the formation of a new Benioff zone. III.- Schematics of the geometry of main elements at the time when the Pucallpa alkalic complex was implaced near surface, i.e., 4.4-5.4 my ago (Steward, 1971). At this stage, the detachment zone was possibly fully developed, and the old Benioff zone was in process of being extinct. IV.- Present structure of the subduction zone. This figure is the same of Fig. 10 with the Pucallpa alkalic complex and Hussong's et al. (loc, cit.) oceanic underthrust fault added.

Summary and Conclusions

The interest of the earth scientists to solve the Nazca-South American plates dynamic puzzle have been remarkably high. In this paper the most important contributions to the understanding of plate dynamics in collision boundaries have been reviewed. Results from marine geophysical research on the Nazca Plate will be published in a special memoir (L. Kulm, personal communication).

The following are some of the important finding and conclusions for the Peru-North Chile segment of the Nazca-South American plates subduction zone:

i) From controlled source seismology research: A low dipping angle, about 6° of the crust-upper mantle interface, towards the continent off south and central Peru; and underthrust faulting on the oceanic crust in deep ocean.

ii) The gravitational anomalies reduced to a sea level datum reflect the structural complexity and the linear pattern of the major tectonic elements : Peru-trench, Andean tectonic framework, and the major, approximately, E-W tectonic deflections: Abancay-Pisco and Huancabamba-Paita.

iii) The 64-min geomagnetic co-transfer function correlates well with the geologic and tectonic pattern in southern Peru, in particular, with the Abancay-Pisco deflection. In the 16-min period data, however, the correlation is less impressive, but it shows a well developed anomaly on the land extension of the oceanic Nazca ridge. It is proposed that this anomaly is associated with ridge extension under the continent.

iv) A model is proposed to explain the present seismic activity distribution, petrological and geochemical data, the difference in morphology and tectonic pattern between central and southern Peru. The model consists of two normal dipping and different age Benioff zones connected by a detachment zone. This zone affects the upper-mantle and crust. The kinematics history of the Andean subduction zone is also modeled as the result of an alternate succession of the South American plate motion excited by the reactivation of the South Mid Atlantic ridge spreading center. The model does not require that the subducted slab be continuous through the subduction zone nor need to change its dip with depth. The basic differences between the Hasegawa and Sacks (1979) model and the proposed model is the absence of the subducted slab between the western shallow seismicity activity belt and the most eastern intermediate depth belt. Future efforts should be devoted to obtain data with the objective of resolving these differences.

References

Aldrich, L.T., M. Casaverde, R. Salgueiro , J. Bannister, F. Volponi, S. del Pozo, L. Tamayo, L. Beach, D. Rubin, R. Quiroga, and E. Triep, Electrical conductivity studies in the Andean cordillera, CIW Year Book 71, 317-320, 1972.

Barazangi, M., B.L. Isacks, Spatial distribution of earthquakes and subduction of the Nazca plate beneath South America, Geology, Vol. 4, 686-692,1976.

Barazangi, M., B.L. Isacks, Subduction of the Nazca plate beneath Peru: evidence from spatial distribution of earthquakes, Geophys. J.R.Astr.Soc., 57, 537-555, 1979.

Breville, B.L., C.W. Beirle, J.R. Sanders, J.P. Voss, and L.E. Wilcox, Bouguer gravity map of South America, Technical paper, N73-2, DMAAC, St. Louis, Mo., 1973.

Casaverde, M., A.A. Giesecke, R. Salgueiro, S. del Pozo, L. Tamayo and L.T. Aldrich, Conductivity anomaly under the Andes, Geofis. Inst., Vol. 8, 55-61, 1968.

Hasegawa, A., and I.S. Sacks, Subduction of the Nazca plate beneath Peru as determined from seismic observations, CIW Year Book 78, 276-284, 1979.

Hayes, D.E., A geophysical investigation of the Peru-Chile trench: Marine Geology, Vol. 4, 309-351, 1966.

Hussong, D.M., P.B. Edwards, S.H. Johnson,

J.F. Campbell, and G.H. Sutton, Crustal structure of the Peru-Chile trench: 8°-12°S latitude, in the Geophysics of the Pacific ocean basin and its margin, AGU, Geophysical Monograph series 19, edited by G.H. Sutton, M.H. Manghnani, and R. Moberly, 71-86, 1976.

James, D.E., Magmatic and seismic evidence for subduction of the Nazca plate beneath central Peru, CIW Year Book 76, 830-831, 1977.

Keller, B., B.T.R. Lewis, C. Meeder, C.Helsley, and R.P. Meyer, Explosion seismology studies of active and passive continental margins, AAPG Memoir 29, 443-451, 1978.

Lamb, H., Hydrodynamics, Dover Pub., 738,pp. 1932 Edition.

LePichon, X., Sea-floor spreading and continental drift, JGR 73 (12), 3661-3697,1968.

Meeder, C.A., B.T.R. Lewis, and J.McClain, The structure of the oceanic crust off southern Peru determined from and ocean bottom seismometer, Earth and Planetary Science Letters 36, 13-28, 1977.

Ocola, L., Seismic hazard evaluation and development of zoning maps in subduction zones, (abstract), XII Symposium of Mathematical Geophysics sponsored by the IUGG Inter Association Committee in Mathetatical Geophysics, 1978

Ocola, L., R.P.Meyer, and L.T. Aldrich, Gross crustal structure under Peru-Bolivia altiplano, Earthquake Notes, Vol. 3-4, 33-48, 1971.

Okada, H., Geophysical implications of the phase ScS on the dipping lithosphere underthrusting western South America,

CIW Year Book 73, 1032-1039, 1974.

Pitcher, W.S., The anatomy of a batholith, J. of the Geol. Soc., Vol. 135, Part 2 , 157-182, 1978.

Rodríguez, R., E.R.Cabré, S.J., and A.Mercado, Geometry of the Nazca Plate and its geodynamic implications, AGU Monograph 19, Edited by G.H. Sutton, M.H. Manghnani, and R. Moberly, 87-103, 1976.

Shepherd, G.L., Shallow crustal structure and marine geology of a convergence zone northwest Peru and southwest Ecuador, Ph. D. Thesis, University of Hawaii, 201 pp. 1979.

Snoke, J.A., and I.S.Sacks, Determination of the subducting lithosphere boundary by use of converted phases, CIW Year Book 74, 266-273, 1975.

Snoke, J.A., I.S.Sacks, and H. Okada, Determination of the subducting lithosphere boundary by use of converted phases. BSSA 67(4), 1051-1060, 1977.

Stauder, W., Subduction of the Nazca Plate under Peru as evidenced by focal mechanism and by seismicity, JGR, Nº180, 1053-1064, 1975.

Steward, J.W., Neogene peralkaline igneous activity in eastern Peru, Bull. Geol.Soc. Amer. 82, 2307-2312, 1971.

Whitsett, R.M., Gravity measurements and their structural implications for the continental margin of southern Peru, Ph. D. Thesis, Oregon State University,1975.

Woollard, G.P., and L. Ocola, Tectonic pattern of the Nazca Plate, IPGH Comisión de Geofísica, Vol. 2(2), 125-149, 1973.

KINEMATICS OF THE SOUTH AMERICAN SUBDUCTION ZONE
FROM GLOBAL PLATE RECONSTRUCTIONS

Rex H. Pilger, Jr.

Department of Geology, Louisiana State University
Baton Rouge, Louisiana 70803

Abstract. Global plate reconstructions have been derived for the Nazca and South American plates for sixteen discrete magnetic anomalies of Late Cretaceous (anomaly 31) and younger age. Kinematic inferences from the reconstructions suggest major reorientations of Nazca-South America plate motions near anomaly 7 (25 Ma) and 21 (48 Ma) times. Since anomaly 7 time plate convergence has been essentially east-west at a fairly high rate. Between anomaly 7 and 21 times, the reconstructions suggest more nearly northeast-southwest convergence, at slight lower rates, except for a period of apparently high rates between anomalies 13 (37 Ma) and 18 (42 Ma).

Plate motions prior to anomaly 21 suggest large components of right-lateral shear, even oblique extension, across the Peru-Chile trench, contradicting geologic evidence in the Andes for subduction in the early Cenozoic. The pre-anomaly 21 reconstructions are probably in error for this reason, as well as for inconsistencies in reconstructed paleomagnetic data noted by other workers. Nevertheless, it is noteworthy that the circa anomaly 21 kinematic reorientation corresponds with the inferred kinematic age of the bend in the Hawaiian-Emperor island-seamount chain.

Introduction

Plate tectonic theory provides a powerful framework for interpretation of mountain belts such as the South American Andes (Fig. 1). The Andes have in fact become a primary focus for study as the contemporary type-example of oceanic subduction beneath a continental plate. An understanding of recent Andean evolution as a convergent plate boundary, based on instantaneous plate motion models, seismology, and volcanology, provides insight into not only the early history of the Andes, but also the history of other inferred continent-ocean subduction systems. The Andean "model" was, for example, first outlined as an analog for the Late Mesozoic of western North America [Hamilton, 1969].

With the progressive increase in knowledge of evolution of the ocean basins, plate tectonics provides an additional, very powerful predictive tool for interpretation of the history of convergent plate margins: global plate reconstructions. By appropriate combinations of marine-derived plate reconstructions, it is possible to produce indirect estimates of relative plate positions for discrete times. With the addition of simplifying assumptions, the discrete reconstructions can be interpolated to provide a kinematic history of the convergent plate boundary.

Recently, Pilger [1981] has derived global reconstructions of South America and the Nazca plate for selected magnetic anomalies from 16 to 3. The reconstructions were one aspect of a study of the subduction of aseismic ridges beneath South America. This study is an extension of the earlier work in that reconstructions for pre-anomaly 16 times are presented and kinematic inferences are included.

Global Plate Reconstruction Methods

Two-plate reconstructions

The most precise plate reconstructions are derived by fitting of corresponding marine magnetic anomalies and linking fracture zones on opposite sides of a spreading center. By assuming that the paleoridge shape is rigidly "frozen-in" and preserved in marine magnetic anomalies, corresponding anomalies (and the paleo-fracture zone segments which link the anomalies) on either side of the spreading center should be congruent.

Unfortunately, no adequate formalism exists for independently predicting the surface configuration or divergent plate boundaries. As a consequence, objective, physically meaningful criteria for fitting corresponding magnetic anomaly identifications (that is, applying the congruency assumption) are absent. Anomaly identifications are semi-discrete data the points, sampling a presumably curvilinear feature - the paleoridge. There is an empirically observed

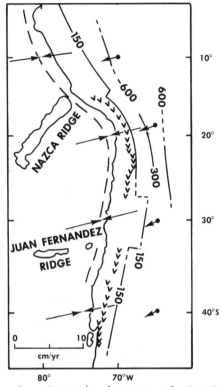

Fig. 1. Tectonic features of the South American subduction zone: Contours on inclined seismic zone (km) and active volcanoes from Barazangi and Isacks [1976]; relative convergence (double-headed arrows) and absolute motion (single-headed arrows) vectors from Chase [1978].

segments of inferred paleoridges, represented by magnetic anomaly identifications, to produce "smooth" curves. As a consequence of this approach, the total rotation poles for a sequence of reconstructions in an ocean do not necessarily correspond, nor is there a compelling dynamic reason why they should. Finite difference parameters approximate the actual kinematics of plate separation [Pitman and Talwani, 1972] and provide indications of changes in plate motion over time intervals between the reconstructions. Most subsequent workers have applied the same reconstruction technique, as above. Exceptions include Francheteau [1973], who assumed a constant spreading pole for a number of anomalies, varying only the rotation angle, and Sclater and Fisher [1974], who included a constant finite difference pole as part of a total rotation reconstruction. Such approaches are somewhat ad hoc, but necessary in those cases in which adequate anomaly identifications are not available.

Pilger [1978] developed a magnetic anomaly-fracture zone fitting algorithm which replaces the trial-and-error approach with an iterative, non-linear least-squares method. The method uses a convenient, although still subjective, fitting criterion designed to produce "smooth" curvilinear paleo-ridge and fracture zone segments.

Srivastava [1978] incorporated an instantaneous velocity closure constraint into trial-and-error reconstructions across a triple-plate junction. Unfortunately, the region he studied (the North Atlantic) involves a triple-spreading center, three-plate system which existed for only a very short time, as appears to be typical of other such systems. Therefore his approach is of limited applicability.

The reconstruction methods described above do not readily permit the objective estimation of errors in the parameters which describe the reconstructions. Further, combinations of these reconstruction parameters, the goal of this paper, are nonlinear in the parameters. Consequently, the nature of a compact method of analysis of the propagation of errors is not clear, as the problem is presently formulated.

The Reconstruction Path

The desired goal of reconstructions discussed here, the relative positions of the South American and Nazca plates through time, must be indirectly pursued since the two plates are nowhere separated by a single spreading center manifested by marine magnetic anomalies. Rather, several spreading centers separate the two plates, providing a reconstruction path of combinations of available two-plate reconstructions.

Global "instantaneous" plate motion models already exist [Minster and Jordan, 1978; Chase, 1978] which probably provide an adequate approximation of plate motions for the past five million years (Fig. 1). Since finite rotation parameters are not vectors, changes in plate

tendency for ridge segments to be orthogonal to non-transform faults, but there are ample examples of non-orthogonality to discourage application of the observation (as a fitting criterion) to plate reconstructions. Transform faults presumably trend parallel with the instantaneous direction of seafloor spreading, along small circles centered around the pole of rotation. However, changes in the direction seafloor spreading result in overprinting of on transform directions and probably contribute to the observed complexity of many fracture zones. As a result, a small circle criterion probably cannot be readily used as a constraint on plate reconstructions. In any case, the total rotation pole will, all probability, differ from the instantaneous motion pole.

As a consequence of the problems in developing fitting criteria, most workers have used subjectiv, trial-and-error techniques in fitting correspondine magnetic anomalies to produce relative plate reconstructions. McKenzie and Sclater [1971], Pitman and Talwani [1972], and Weissel and Hayes [1972] derived plate reconstructions by aligning corresponding fracture zone crossings (where possible) and corresponding motion parameters are required for earlier periods. Pre-five million year reconstructions are poorly approximated by extrapolation of instantaneous plate motion models. For earlier times, finite plate reconstructions are available for various marine magnetic anomalies across the spreading centers which, in effect, separate the South American and Nazca plates.

Ladd [1976] has published plate reconstruction parameters between South America and Africa for anomalies 5, 13, 20, 28, and 34. Ladd's evidence suggests no major change in plate motion directions has occurred in the interval between anomaly 5 and 34 times.

Recently, Norton and Sclater [1979] described reconstructions in the Indian Ocean of the African, Indian and Antarctic plates for anomalies 16, 22, 28, and 34. The anomaly 16 reconstruction is a successor to an earlier anomaly 13 reconstruction published by McKenzie and Sclater [1971]. Few lineations between anomaly 5 and 22 are observed across the Indian-African boundary so that the inferred anomaly 16 age of the reconstruction of Norton and Sclater [1979] is an interpolation of sorts.

Weissel et al. [1977] have presented reconstructions across the southeast Indian Ridge, between Australia and Antartica, for anomalies 5, 6, 8, 12, 13, 18, and closure (assumed to be about anomaly 22 time). Assuming that Australia has moved as part of the Indian plate since initial rifting between Australia and Antarctica allows use of Weissel et al.'s parameters for Indian-Antarctic reconstruction. Additional reconstructions of India-Antarctica have been described by Norton and Sclater [1971] for anomalies 28 and 34.

Recently, Cande et al. [1981] argued that published pre-anomaly 18 identifications between Australia and Antarctica are in error. Consequently, the applicability of the anomaly 22 reconstruction of Weissel et at. [1977] is questionable. Nevertheless, Norton and Sclater [1979] have demonstrated that the anomaly 22 rotation parameters of Weissel and others are compatible with existing magnetic anomaly identifications across that part of the southeast-Indian ridge which separates the Indian and Antarctic plates, west of Ninetyeast Ridge.

Molnar et al. [1975] derived reconstructions across the Antarctic-Pacific plate boundary for magnetic anomalies 5, 6, 13, 25, 31, and 34 (closure). Revised closure parameters have been published by Weissel et al. [1977].

Pilger [1978], using methods discussed above, published reconstructions of the Pacific and Nazca plates for anomalies 7, 8, 9, 10, 11, 12, 13 and 16. For that study, identifications and corresponding fracture zone crossings for anomalies 5 and 6 and those older than 16 were considered inadequate for total rotation reconstructions. A subsequently published bathymetric map, by Mammerickx and Smith [1979], includes additional fracture zone crossings which,

together with magnetic anomaly identifications by Herron [1972] and Weissel et al. [1977], permit derivation of an anomaly 18 reconstruction, described in the next section. Older reconstructions for the Nazca-Pacific plate system have also been newly derived by the methods outlined below.

New Pacific-Nazca Reconstructions

Plate reconstructions as old as anomaly 34 appear to be possible around the circuit of South America-Africa-India-Antarctica-Pacific. However, reconstructions of the Pacific and Nazca plates older than anomaly 16 have not been published. This gap is not surprising due to (1) the sparse ship track coverage in the older parts of the Nazca and east-central Pacific plates and (2) the absence of older parts of the Nazca plate, which have been subducted.

Pilger [1978] did not attempt to derive reconstructions older than anomaly 16 for the reasons noted; however, the recently revised bathymetric map of the east-central Pacific [Mammerickx and Smith, 1979] includes new fracture zone crossings which are located adjacent to anomaly 18 identifications on the Nazca plate. The new and previous fracture zone crossings and anomaly identifications on the Nazca and Pacific plates are adequate for an anomaly 18 reconstruction, using the fitting method of Pilger [1978]. The parameters of the anomaly 18 reconstruction (Fig. 2) are listed in Table 1.

Data are still inadequate for total rotation anomaly-fitting reconstructions for anomalies older than 18. However, it is possible to derive reconstructions of the positions of segments of the Pacific-Farallon (Nazca) spreading center for anomalies 25 and 32 relative to their anomaly 18 position, based on Pacific plate identifications above. Then, by assuming symmetrical sea floor spreading, the incremental anomaly 18 to 25 and 25 to 32 reconstructions can be combined (doubling the rotation angles) to produce estimates of the total Pacific-Nazca (Farallon) plate reconstructions for anomalies 25 and 32.

Using the anomaly identifications of Herron [1972] and Weissel et al. [1977] and fracture zone crossings of Mammerickx and Smith [1979], Mammerickx et al. [1975], and Molnar et al. [1975], Pacific plate anomaly 18 identification were fit to anomaly 25 crossings, and anomaly 25 to anomaly 32. The parameters of these reconstructions (Fig. 3) are listed in Table 1, as well as the total rotation parameters for anomalies 25 and 32, derived by combining the anomaly 18, 18-25, and 25-32 recon-structions, with the finite difference rotation angle doubled (Fig. 2).

Interpolation of Reconstructions

Two-plate reconstruction parameters exist within the South American-Nazca plate circuit for anomalies 5, 6, 7, 8, 9, 10, 11, 12, 13, 16, 20,

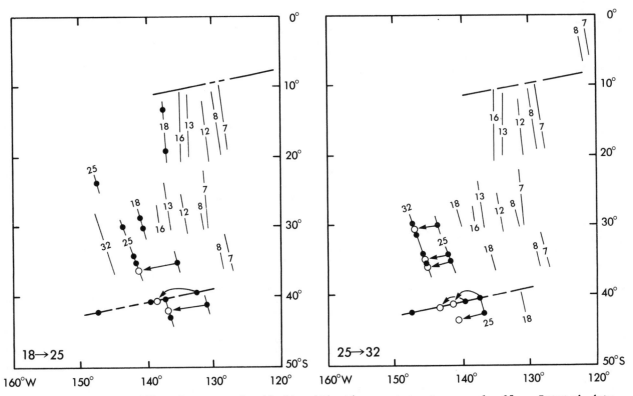

Fig. 2. a. Pacific plate anomaly 18 identifications rotate to anomaly 25. Rotated data points, open circles; unrotated, solid circles. b. Pacific anomaly 25 identifications rotated to anomaly 32 Rotated data points open circles; unrotated, solid circles.

21, 22, 25, 28, 31, 32, and 34. However, reconstructions for all of these anomalies do not exist for every recon-structed plate pair in the circuit. In order to calculate a representative reconstruction sequence, interpolation is necessary.

The simplest interpolation approach involves an assumption of constant spreading rate and pole of rotation between discrete reconstructions. The method restricts changes in plate motion to occur at the times of magnetic anomaly identifications. As such, the method is a source of error. However interpolations of some sort must be used if

adequate reconstruction sequences are to be derived.

Other Sources of Error in Plate Reconstructions

The subjectivity of reconstruction methods introduce errors of uncertain magnitude. These errors are compounded when combined with the finite variable widths of and errors in anomaly and fracture zone identifications.

Errors due to interpolations are of two kinds. The instantaneous pole of motion and the apparent rate of motion may be continuously

Table 1. Pacific-Nazca-Paleoridge Reconstruction Parameters

Fixed Plate Anomaly #	Rotated Plate Anomaly #	Rotation Pole Latitude, deg.	Longitude, deg.	Rotation Angle*, deg.
A. PCFC-18	NAZC-18	72.8	-112.6	-57.1
B. PCFC-25	PCFC-18	80.6	132.7	-6.3
C. PCFC-25**	NAZC-25	75.1	-125.5	-68.8
D. PCFC-32	PCFC-25	74.9	151.5	-3.4
E. PCFC-32***	NAZC-32	75.6	-133.0	-75.3

*Positive counter-clockwise. **Combination of B (with doubled rotation angle) and A. ***Combination of D (with doubled rotation angle) and C.

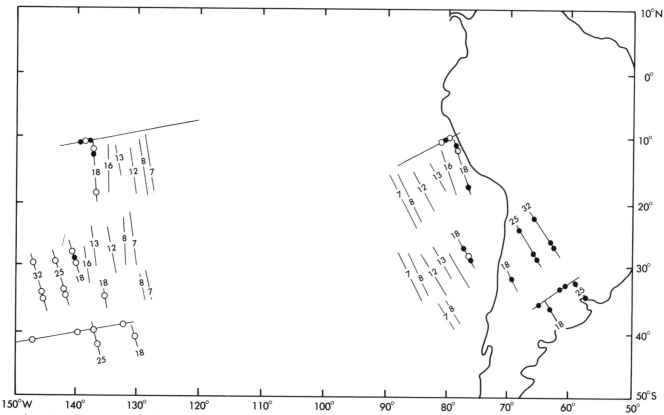

Fig. 3. Nazca-Pacific reconstructions for anomalies 18, 25, and 32. Anomaly identifications and fracture zone crossings from Herron [1972], Mammerickx et a;. [1975], Molnar et al. [1975], Weissel et al. [1977], and Mammerickx and Smith [1975]. Nazca plate identifications and crossings (rotated and unrotated) indicated by solid circles; Pacific by open circles. Anomaly 18 reconstruction derived using Nazca and Pacific plate data and method of Pilger [1978]. Anomaly 25 and 32 reconstructions obtained by combination of anomaly 18 reconstruction with 18-25 and 25-32 finite difference reconstructions (Fig. 2), assuming symmetrical seafloor spreading.

changing. In addition, uncertainties in the geomagnetic time scale produce probable errors, since estimated spreading rates obviously depend on the time scale.

Composing and Interpreting Global Reconstructions

Choices of magnetic anomalies for composing the global reconstructions have been made based on the availability of reconstructions across individual plate boundaries and a desire for a representative sequence of reconstructions. Interpolations are required for every reconstruction presented here, except anomaly 3 (Table 2). The anomaly 13 reconstruction is probably the best constrained (except for anomaly 3) in that interpolations are required for only one plate pair, Africa-India.

The resulting global reconstructions from the plate pair combinations provide a deceptively detailed sequence of reconstructions. Since the evaluation of errors is a quantitative impossibil-

ity with present techniques, a subjective, inductive approach has been taken.

With a basic assumption of relative constancy of plate motion, perturbations in a global set of reconstructions could reflect either errors or true changes in plate motion. By analyzing each plate pair reconstruction used in the global reconstructions, it should be possible to decide whether a major apparent change is distributed through the reconstruction sequence or recorded in only one plate pair.

If plate motions relative to the earth's interior are largely a product of plate boundary induced forces [Solomon and Sleep, 1974; Forsythe and Uyeda, 1976], then changes in motion between two plates should be recorded most clearly across the boundaries with adjacent plates. If, on the other hand, a perturbation in the global reconstruction corresponds most directly with a plate motion change occurring at a non-adjacent plate boundary, the validity of the reconstructions itself might be called into question, or global plate motions are more closely

Table 2. Plate Pair Reconstruction Parameters

Anomaly Number	Rotation Latitude, deg. N	Pole Longitude, deg. E	Rotation Angle, deg.	Source
Nazca to Pacific				
3	50.9	-87.0	-7.04	a
5	58.6	-89.0	-14.8	i
6	62.2	-90.3	-30.5	i
7	63.0	-90.6	-39.6	b
12	67.1	-99.9	-46.9	b
13	69.4	-102.5	-49.8	b
16	70.8	-107.8	-52.7	b
18	72.8	-112.6	-57.1	c
20	73.4	-115.2	-59.4	i
21	74.0	-118.4	-62.3	i
22	74.4	-120.5	-64.1	i
25	75.1	-125.5	-68.8	c
28	75.4	-130.0	-72.6	i
31	75.5	-131.9	-74.3	i
32	75.6	-133.0	-75.3	c
Pacific to Antarctic				
3	66.2	-83.5	4.20	a
5	68.7	-79.7	8.9	d
6	72.0	-72.0	15.7	d
7	73.5	-65.2	20.0	i
8	73.7	-63.7	21.2	i
12	74.4	-59.0	25.6	i
13	74.7	-57.0	27.9	d
16	75.1	-52.4	30.5	i
18	75.3	-48.5	33.0	d
20	74.8	-49.7	34.8	i
21	74.2	-50.1	36.9	i
22	73.8	-51.6	38.3	i
25	73.0	-53.	41.8	d
28	70.4	-55.8	47.0	i
31	69.5	-56.7	49.4	d
32	69.6	56.0	51.8	i
34	70.0	52.5	67.0	e
Antarctic to India				
3	17.4	32.1	2.84	a
5	10.7	31.6	6.2	e
6	13.8	34.6	12.0	e
7	14.2	33.5	15.1	i
8	14.2	33.2	16.0	e
12	12.5	34.4	18.9	e
13	11.9	34.4	20.5	e
16	11.0	34.6	22.1	i
18	10.3	34.8	23.6	e
20	10.9	33.2	26.0	i
21	11.6	31.7	28.9	i
22	11.9	30.8	30.9	e
25	14.8	21.2	35.3	i
28	17.4	10.9	42.8	f
31	16.5	10.6	47.3	i
32	16.1	10.4	49.9	i
34	14.1	9.7	66.2	f

Table 2 (continued). Plate Pair Reconstruction Parameters

Anomaly Number	Rotation Latitude, deg. N	Pole Longitude, deg.E	Rotation Angle, deg.	Source
		India to Africa		
3	16.0	48.3	-2.84	a
5	14.5	51.6	-5.5	i
6	13.7	53.3	-10.9	i
7	13.5	53.7	-13.9	i
8	13.5	53.7	-14.8	i
12	13.4	54.0	-18.0	i
13	13.3	54.0	-19.6	i
16	13.3	54.1	-20.8	f
18	13.2	51.8	-22.6	i
20	13.1	49.2	-25.0	i
21	13.0	46.7	-28.1	i
22	12.9	45.3	-30.1	f
25	15.9	36.5	-33.1	i
28	18.8	26.2	-38.4	f
31	18.8	26.1	-41.8	i
32	18.B	26.0	-43.8	i
34	18.7	25.8	-56.0	f
		Africa to South America		
3	63.9	-34.3	-1.5	a
5	70.0	-35.0	-3.8	g
6	61.8	-35.0	-7.4	i
7	59.9	-35.0	-9.6	i
8	59.5	-35.0	-10.2	i
12	58.4	-35.0	-12.3	i
13	59.9	-35.0	-13.5	g
16	61.3	-35.3	-14.9	i
18	63.0	-35.6	-16.2	i
20	63.0	-36.0	-18.0	g
21	63.0	-36.0	-19.1	i
22	63.0	-36.0	-19.9	i
25	63.0	-36.0	-21.7	i
28	63.0	-36.0	-24.3	g
31	63.0	-36.0	-26.1	i
32	63.0	-36.0	-27.2	i
34	63.0	-36.0	-33.8	g

Sources: a, Chase [1978], b, Pilger [1978]. c, This paper (Table 1). d, Molnar et al. [1975]. e, Weissel et al. [1977]. f, Norton and Sclater [1979]. g, Ladd [1976]. i, interpolated.

interdepedent than the dynamic assumption suggests.

Nazca–South American Global Plate Reconstruction

From the reconstructions discussed above, and presented in Table 2, Nazca-South American reconstructions have been derived (Table 3) by the sequence of rotations Nazca-Pacific-Antarctic-Indian-African-South American, for anomalies 3 (4.315 Ma), 5 (9.55), 6 (19.96), 7 (26.005), 8 (27.77), 12 (33.82), 13 (37.08), 16 (39.365), 18 (41.555), 20 (44.485), 21 (48.06), 22 (50.42), 25 (56.21), 28 (64.56), 31 (68.345), and 32 (70.51). Anomaly ages apply to midpoints of anomalies in the time scale of Lowrie and Alvarez [1981].

From the reconstructions, illustrated in Figure 4, several kinematic observations are apparent. The most significant involves recognition of changes in plate motion (assuming that the reconstructions are representative of a 'smooth' kinematic pattern) near anomalies 7 and 21. Prior to anomaly 21, the reconstructions suggest that

Table 3. Nazca – South America Plate Reconstruction and Kinematic Parameters

Anomaly Number	Total Rotation			Finite Difference			
	Pole Latitude, deg.	Pole Longitude, deg.	Angle	Pole Latitude, deg.	Pole Longitude, deg.	Angle	Rate deg./my
				48.9	-86.4	3.83	.88
3	48.9	-86.4	-3.83				
				65.4	128.3	5.6	1.06
3	60.9	-105.7	-9.1				
				60.9	-93.0	1.3	1.08
6	60.8	-98.7	-20.4				
				60.1	-95.8	6.1	1.01
7	60.0	-98.4	-26.5				
				42.3	171.8	1.4	0.80
8	62.0	-102.2	-27.3				
				51.0	-166.0	4.2	0.69
12	63.6	-111.4	-30.8				
				49.3	150.1	2.4	0.74
13	66.1	-116.9	-32.4				
				31.8	166.9	3.2	1.38
16	65.3	-125.9	-34.7				
				65.6	174.7	4.1	1.89
18	67.0	-130.4	-38.5				
				52.8	152.0	2.4	0.81
20	68.3	-135.0	-40.4				
				39.1	153.6	1.8	0.49
21	68.8	-139.7	-41.6				
				39.4	155.4	1.1	0.48
22	69.1	-141.7	-42.4				
				25.6	164.9	3.3	0.57
25	68.5	-152.2	-44.5				
				-15.6	160.8	4.4	0.53
28	65.7	-165.2	-44.7				
				2.7	167.0	2.7	0.71
31	63.7	-170.9	-45.8				
				-42.8	167.3	1.5	0.67
32	62.0	-172.2	-45.4				

the Farallon (Nazca) plate was moving slowly north-northwest with respect to South America. After anomaly 21, Farallon-South American convergence began in a northeast-southwest direction. At about anomaly 7 time, convergence became more nearly east-west. Perturbations of plate motion direction are suggested by the relative positions of the anomaly 5, 16, and 32 reconstructions.

Plate Kinematics from Discrete Reconstructions

Pitman and Talwani [1972] and Weissel and Hayes [1972] were among the first to separate the plate reconstruction problem from the derivative problem of plate kinematics, using actual reconstructions. Both pairs of workers derived finite difference rotations between total rotations (which describe the reconstruction) and used the derived parameters to predict flow-lines and to synthesize paleo-transform faults for comparison with actual mapped fracture zones. The former authors showed that the derived synthetic faults matched actual fracture zones fairly well, providing support for both the reconstructions and for the assumption that the reconstructions were closely enough spaced to adequate record the kinematic pattern.

Using finite difference parameters between reconstructions presented here (Table 3), instantaneous motion vectors have been calculated for each interval (Fig. 5). It is apparent that the two major changes in plate motion direction coincide with changes in rates of Nazca-South American convergence. Further, the anomaly 16 perturbation also coincides with a period of anomalously high calculated convergence rates. The anomaly 5 perturbation is not obvious in the rates, however.

It is quite desirable to establish whether either the major plate motion changes or the perturbations truly reflect changes in relative motion of two adjacent plates in the reconstruction circuit or are cumulative results of small changes or errors in calculated motions in the circuit. In order to address this question, the

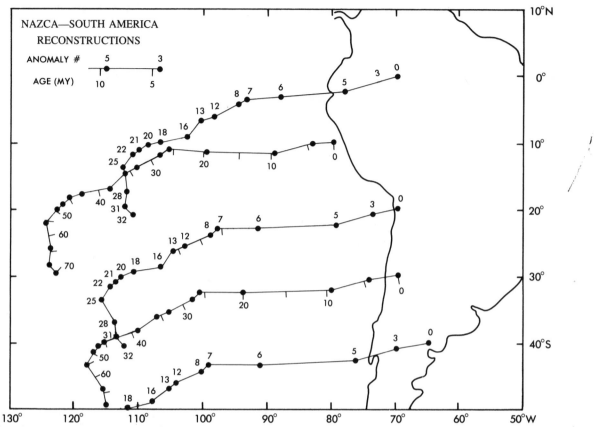

Fig. 4. Global reconstructions of arbitrary points on Nazca (Farallon) plate relative to fixed South America, using parameters in Table 3. Anomaly numbers are above reconstructed loci, corresponding with solid circles. Time scale (ma) below loci, according to Lowrie and Alvarez [1981].

relative reconstructions of individual plate pairs have been applied to points proximal to the Nazca-South American plate boundary. In this way, the importance of changes in relative motions among the plate pairs to the global reconstructions can be assessed.

By portraying the plate-pair reconstructions in the vicinity of South America, important changes in motion, pertinent to the global reconstructions of Nazca-South America, are highlighted. A subtle change in plate motion of the India-Antarctic pair, for example, may be more pronounced when integrated into the global circuit for a distant plate pair such as Nazca-South America. However, it is important to recognize that the plate pair reconstructions, as portrayed in Figure 6, are semi-artificial in the South American frame of reference. The position of the reconstructed points for each anomaly relative to other points reconstructed for other anomalies is significant. However, position of the reconstructed points relative to South America is not meaningful.

From comparison of the global reconstructions (Fig. 4) with the plate-pair reconstructions (Fig.

6) it is apparent that the circa anomaly 7 Nazca-South American motion change most directly corresponds with a motion change involving the Pacific-Nazca plate pair. The anomaly 7 event records the break-up of the Farallon plate to form the Nazca and Cocos plates. Shortly after anomaly 7 time, the newly formed Nazca plate began moving in a more easterly direction, relative to the Pacific plate, than previously [Handschumacher, 1976].

The other major change in Nazca-South America motion directions, at about anomaly 21 time (Fig. 4), is not obvious in the motion directions of either the Nazca or South American plate relative to other adjacent plates. There is a significant change in inferred Pacific-Nazca spreading rates at about this time, between anomaly 18 and 32 - note that the anomaly 21, 22, 28 and 31 Pacific-Nazca reconstructions are interpolated. Data are inadequate to indicate changes in spreading rate in the South Atlantic at anomaly 21 time. In any case, a major assumption of the geomagnetic time scale is essential constancy of South America-Africa spreading rate between control points [Lowrie and Alvarez, 1981]. The circa anomaly 21 change in Nazca-South American plate motions

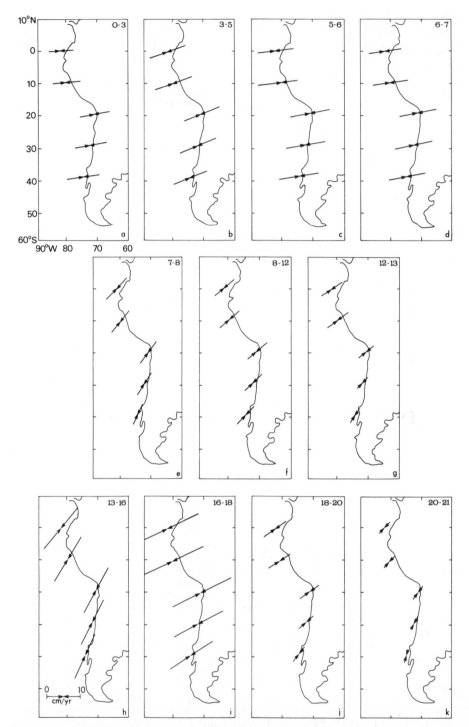

Fig. 5. Instantaneous convergence vectors for reconstruction intervals, assuming constant finite difference poles and rates, between reconstructions for anomalies 3 through 21 (Table 3).

corresponds directly with the changes in motion direction and rate of India-Antarctica and motion rate of India-Africa (Fig. 6). Since neither the South American or Nazca plates share a boundary with the Indian plate (except at one point for the former - the South Atlantic triple plate junction), the basic dynamic assumption (dominant driving forces are at plate margins) casts doubt on the pre-anomaly 21 reconstructions. That is, the apparent lack of significant changes

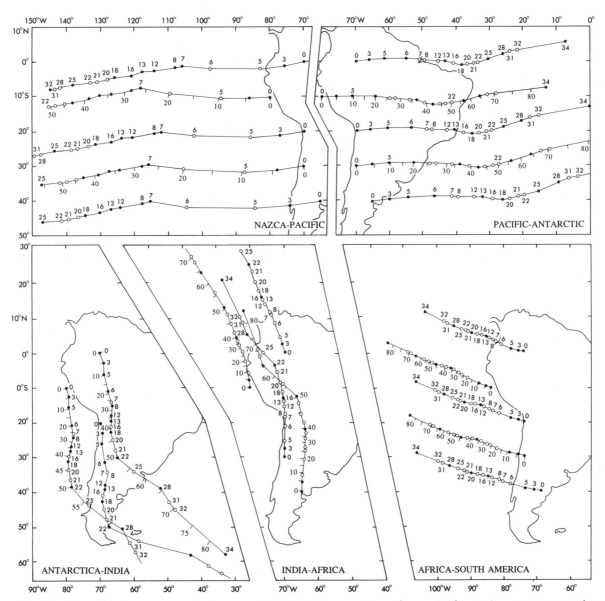

Fig. 6. Plate pair reconstructions used in global reconstructions of Figure 4. Reconstructions are fixed relative to second plate and South America in present coordinates. Interpolated reconstructions are indicated by open circles. a, Nazca to Pacific. b, Pacific to Antarctic. c, Antarctic to Indian. d, Indian to African. e, African to South American.

in motion between the Nazca and South American and adjacent plates near anomaly 21 (allowing for constant motion interpolations) implies either an erroneous pre-anomaly 21 plate pair reconstructions or an erroneous dynamic assumption.

In support of the essential validity of the reconstructions (and consequent error in the dynamic assumption) note that the anomaly 21 motion change (Fig. 4) is close to inferred kinematic age of the bend (between anomalies 23 and 21) in the Hawaiian-Emperor island-seamount

chain [Pilger and Handschumacher, 1981; Butler and Coney, 1981], and a change in Pacific-Farallon plate motion observed in magnetic anomaly and fracture zone trends in the North Pacific [Atwater and Menard, 1970]. The latter change is not obvious in magnetic anomaly identifications in the South Pacific, however. Arguing against the validity of the pre-anomaly 21 reconstructions are discrepancies in early Cenozoic and late Mesozoic paleomagnetic data between the Pacific and other plates [Suarez and Molnar, 1980; Gordon and Cox, 1980]. These discrepancies seem to require an additional plate boundary in the South Pacific or Antarctica. Even more critical is the direction of Nazca-(Farallon) South American plate motion implied by the anomaly 32 to 21 reconstructions. The reconstructions indicate dominantly shear motion, with a slight extensional component across the Chilean portion of the Peru-Chile trench. Yet, geological evidence indicates volcanic arc-like igneous activity was occurring in the early Cenozoic of the central Andes (Pilger, 1981), implying subduction (therefore plate convergence) of an oceanic plate beneath South America. Either an additional oceanic plate is required between the Farallon and South American plates, or the reconstructions are in error.

It is not possible to better resolve the problem of the validity of the pre-anomaly 21 Nazca-South American reconstructions from present data. In the future paleomagnetic studies would appear to hold the most promise for confirming or discounting the general reconstruction pattern indicated in Figure 4.

The anomaly 16 perturbation in the global reconstructions is readily seen to correspond with an apparent pulse of rapid seafloor spreading between anomalies 13 and 21 across the Farallon-Pacific plate boundary (Figure 4). It is not at all clear whether this pulse is real, or a result of erroneous magnetic anomaly identifications or errors in the geomagnetic time scale. The last possibility seems unlikely since no such pulse is apparent across the other plate boundaries where adequate reconstructions exist, particularly for anomalies 13 to 18 between Australia and Antarctica.

The other slight perturbations in plate motions inferred from the global plate reconstructions of Figure 4 are probably better attributed to errors in the plate-pair reconstructions and in the interpolation methods than to actual motion changes. Similarly the fluctuations in inferred rates of motion from anomaly to anomaly cannot be accepted without reservations.

Summary and Conclusions

It is fair to conclude that two well-established major trends in Nazca-South American plate motion are indicated by the global plate reconstructions. That is, Farallon (Nazca)-South American plate convergence between circa anomaly 21 and 7 times was in an eastnortheast-westsouthwest direction. Shortly after anomaly 7 time, the convergence direction changed to more nearly east-west, persisting to the present.

Convergence rates, estimated from the reconstructions, have been high over much of the entire period since anomaly 21 time, generally exceeding 8 cm/yr. Apparent convergence rates as high as 20 cm/yr have been calculated, but their validity is questionable.

The pre-anomaly 21 reconstructions are also of doubtful validity, for several reasons. Future work will be essential in establishing viable

early Cenozoic and Late Cretaceous reconstructions.

The reconstructions (Fig. 4) and the instantaneous vectors (Fig. 5) (derived from the finite difference rotations) provide a tentative framework for interpreting the Cenozoic history of the Andean subduction zone. Of particularly significance is the post-anomaly 7 (25 Ma) plate motion change, with its implications for changing convergence rate normal to and parallel with the trench.

References

Atwater, T., and H. W. Menard, Magnetic lineations in the Northeast Pacific, Earth Planet. Sci. Lett. 7, 445-450, 1970.

Barazangi, M., and B. L. Isacks, Spatial distribution of earthquakes and subduction of the Nazca plate beneath South America, Geology, 4, 686-692, 1976.

Butler, R. F., and P. J. Coney, A revised magnetic polarity time scale for the Paleocene and early Eocene and implication for Pacific plate motion: Geophys. Res. Lett., 8, 304-304, 1981.

Cande, S. C., Mutter, J., and J. K. Weissel, A revised model for the break-up of Australia and Antarctica: Am. Geophys. Union, Spring Ann. Meeting, Baltimore, Eos, 62, 384, 1981.

Chase, C. G., Plate kinematics: the Americas, east Africa, and the rest of the world, Earth Planet. Sci. Lett., 37, 355-368, 1978.

Forsyth, D., and S. Uyeda, On the relative impor tance of the driving forces of plate motion, Geophys. Jour. Royal Astron. Soc., 43, 163-200, 1975.

Francheteau, J., Plate tectonics model of the opening of the Atlantic Ocean south of the Azores, in Implications of Continental Drift to the Earth Sciences, S. K. Runcorn and D. H. Tarling, ed., Academic Press, New York, 197-202, 1973.

Gordon, R. G., and A. Cox, Paleomagnetic test of the early Tertiary plate circuit between the Pacific basin plates and the Indian plate: Jour. Geophys. Res., 85, p. 6534-6546, 1980.

Hamilton, W., The volcanic central Andes--a modern model for the Cretaceous batholith and tectonics of western North America, Oregon Dept. Geology and Mineral Industries, 65, 175-184, 1969.

Handschumacher, D. W., Post-Eocene plate tectonics of the eastern Pacific, Am. Geophys. Union Mon. 19, 177-202, 1976.

Herron, E. M., Sea-floor spreading and the Cenozoic history of the East Central Pacific, Geol. Soc. Am. Bull., 83, 1671-1692, 1972.

Ladd, J. W., Relative motion of South America with respect to North America and Caribbean tectonics, Geol. Soc. Am. Bull., 87, 969-976, 1976.

Lowrie, W., and W. Alvarez, One hundred million years of geomagnetic polarity history, Geology, 9, 392-397, 1981.

Mammerickx, J., R. N. Anderson, H. W. Menard, and S. M. Smith, Morphology and tectonic evolution of the East Central Pacific, Geol. Soc. Am. Bull., 86, 111-118, 1975.

Mammerickx, J., and S. M. Smith, Bathymetry of the Southeast Pacific, Geol. Soc. Am. Map and Chart Series MC-26, 1979.

McKenzie, D. P., and J. G. Sclater, The evolution of the Indian Ocean since the Late Cretaceous, Geophys. Jour. Royal Astron. Soc., 25, 437-528, 1971.

Minster, J. B., and T. H. Jordan, Present day plate motions, Jour. Geophys. Res., 83, 5331-5354, 1978.

Molnar, P., T. Atwater, J. Mammerickx, and S. M. Smith, Magnetic anomalies, bathymetry, and the tectonic evolution of the South Pacific since the Late Cretaceous, Geophys. Jour. Royal Astron. Soc., 40, 383-420, 1975.

Norton, I. O., and J. G. Sclater, A model for the evolution of the Indian Ocean and the breakup of Gondwanaland, Jour. Geophys. Res., 84, 6803-6830, 1979.

Pilger, R. H., Jr., A method for finite plate reconstructions, with applications to Pacific-Nazca plate evolution, Geophys. Res. Lett., 5, 469-472, 1978.

Pilger, R. H., Jr., Plate reconstructions, aseismic ridges, and low-angle subduction beneath the Andes, Geol. Soc. Am. Bull., 92, 448-456, 1981.

Pitman, W. E., III, and M. Talwani, Sea-floor spreading in the North Atlantic, Geol. Soc. Am. Bull., 83, 619-646, 1972.

Sclater, J. G., and R. L. Fisher, Evolution of the central Indian Ocean with emphasis on the tectonic setting of the Ninety-east Ridge, Geol. Soc. Am. Bull., 85, 683-702, 1974.

Solomon, S. C., and N. H. Sleep, Some simple physical models for absolute plate motions, Jour. Geophys. Res., 79, 2557-2567, 1974.

Srivastava, S. P., Evolution of the Labrador Sea and its bearing on the early evolution of the North Atlantic, Geophys. Jour. Roy. Astron. Soc., 52, 313-357.

Suarez, G., and P. Molnar, Paleomagnetic data and pelagic sediments facies and the motion of the Pacific plate relative to the spin axis since the Late Cretaceous, Jour. Geophys. Res., 85, 5257-5280, 1980.

Weissel, J. K., and D. E. Hayes, Magnetic anomalies in the Southeast Indian Ocean, in Hayes, D. E., ed., Antarctic Oceanology II: The Australia-New Zealand Sector, Am. Geophys. Union, 163-196, 1972.

Weissel, J. K., D. E. Hayes, and E. M. Herron, Plate tectonics synthesis: the displacements between Australia, New Zealand, and Antarctica since the Late Cretaceous, Mar. Geol., 25, 231-277, 1977.

GEODYNAMICS OF THE ARGENTINE ANDEAN ARC AND RELATED REGIONS

Bruno A. J. Baldis (1) and Jose Febrer (2)

(1) CONICET, Rivadavia 1917, 1033 Bs. As., Argentina
(2) CNIE, Avda. Mitre 3100, 1663 San Miguel, Argentina

Argentine participation in the work of WG2 has been concerned chiefly with the effects of the "rearguard action" of the Nazca Plate subduction zone since the country contains the western flank of the southern Andean Arc. For that reason, the advances in knowledge of the Geodynamics of the Andean Arc and the Scotia Arc are presented in two parts, one concerned with geotectonic studies and the other with advances in geophysics and the interpretation of the position of the asthenosphere at the base of the American Plate.

Part One: Advances in Geotectonic Knowledge

A. Plan for the integrated geotectonic analysis of Argentine territory.

Until 1971 there was no plan in existence in Argentina for the study of the structural geology of its territory; this was only partially known and in detail in only a few areas. Plans for geological studies by means of aerial photograpgs and, from 1973 onwards, by satellite imagery made possible rapid advances in knowledge of the structure of the country.

The plan was initiated by Bruno Baldis in 1973 at the University of Buenos Aires with the collaboration of R. Sarudiansky, continuing in 1974 with J. Saltify and J. Viramonte. An integrated plan for the whole of Argentine territory was established in 1975 and adopted as the principal theme for the work of the Argentine Andean Geotectonic Study Group created with the collaboration of WG2 in 1976. In addition to those named above, the following people participated: A. Otriz, E. Uliarte, A. Vaca and J Zambrano and, at the institutional level, the National University of San Juan. At present this plan is being developed under the auspices of the Argentine National Council for Scientific and Technical Research. As part of the plan for geotectonic analysis a geotraverse of northern Patagonia was made which is discussed below.

Up to the moment, a total of 640,000 km^2 have been studied, comprising the Andean and Pre-Andean zones of Argentina between the parallels of 22°S and 35°S (Fig. 1). A systhesis of this study is presented in Fig. 2, to which has been added complementary information provided by the Direccion de Fabricaciones Militaires and Plan NO Argentina (NOA).

The study carried out the following steps:
1. Compilation of a structural base map at a scale of 1:1,000,000 with the addition of a bibliography, aerial photogeology and satellite imagery with corresponding ground control.
2. The classification of structural types based on their size.
3. Determination of alignments of megafractures.
4. Tentative establishment of the ages of initiation and activity of fractures.
5. Compilation of profiles of medium to great depths including control by earthquake locations with the aim of determining the continuity of superficial structures into the deep crust.
6. Dynamic analysis of the structural systems to learn their present behavior and the origin of the forces involved. The whole study of the northern area was centered on the sub-Andean zone and the Eastern Andes, leaving the area of La Puna de Atacama for a future study.

The analysis then was continued towards the southwest between the Pampean, Calchaqui and Aconquija systems and thence to culminate towards the south in the area contained between the great Sierras Pampeanas de Cordóba and San Luis and the Central Valley of Chile.

The central zone of Argentina is known as the Central Craton and is composed of metamorphic and magmatic rocks which behave rigidly. Geotectonic analysis has established that the Central Craton has at its periphery controlled marine incursions since the Lower Paleozoic, but in its turn has been broken up by a conjugate system of megafractures (Fig. 3) crossing the Andean direction; many of the latter have been active from Lower Precambrian to the present.

Except for the sectors of internal reaction, the whole of the Central Craton appears to exert pushes in a general east to west sense, cumminating in various fronts of fracturing. These pushes have been named "Pampean type" and they form part of the mechanics of the translation of the South American Plate towards the west.

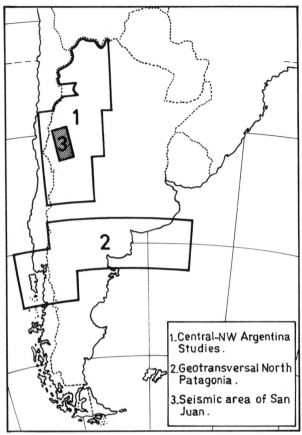

Fig. 1. Regional studies and project developping.

On the Pampean front of fracturing are contraposed from a more westerly direction structural systems inverse in position to its force vectors giving rise to strongly active structures in the Upper Tertiary and Quaternary.

Some of the structures are old, such as the Taconic Front of northwestern Argentina or the Tontal Front in the Argentina PreCordillera formed in the Lower Paleozoic.

This general direction of "counterpush" from the Andean Arc may be interpreted not only as a reaction to the push of the Craton but also as a manifestation of relief with a style antithetic to the subduction plane at the continental margin.

From the adjoining geotectonic maps (Figs. 2,3) it can be observed that the counterpush of the Andes, with a frontal position in the SubAndean Sierras of Perú and Bolivia, on reaching Argentine territory turns back and is located above the Taconic Front of the Cordillera or Eastern Andes remaining in that position as far as the parallels 25°S and 26°S approximately.

The SubAndean System ends at parallel 23°S pivoting over a tectonic node, Valle Grande, which like the fulcrum of a lever holds up the structures, a short arc running in the opposite

direction, towards the west between parallels 23°30´S and 25°S (Santa Barbara System).

South of parallel 26°S the front of Andean counter forces is progressively displaced westward in the form of an arc whose center is the seismically quiet Pipanaco depression. The Andean "counter front" resumes a southern direction where it reaches the PreCordillera, over the Tontal Front of fracturing, which is preceded in its northern part by a belt of overlying Upper Tertiary and Quaternary.

For its part the Pampean Front (of pushing) is displayed in notable form from the eastern flank of the Bolson de Pipanaco and, being displaced progressively toward the west, controls the great zone of fracturing of the Valle Fertil Front and the great, half-covered blocks of the provinces of San Juan and Mendoza.

To sum up: the Pampean forces and the corresponding Andean counterforces give rise to a form of arc and counterarc, abutting against the Massif of La Puna and generating zones of cancellation of forces (Pipanaco), of torsional forces (Node of Valle Grande) and of strong shearing (seismic zone of San Juan).

The whole system is currently active including the megafractures generated at times prior to the beginning of continental drift.

The completion of a study of the relation between the distribution of hypocenters and the geotectonic elements described in this report is expected in the near future.

B. The transcontinental structure of north Patagonia

The study by Turner and Baldis (1978) of the structural relations between the northern limits of Patagonia and the northern part of the South American Plate has made possible the following conclusions:
1. A structural continuity is observed that, starting from the edge of the Argentine continental shelf continues towards the west in the Colorado and Neuquen basins (Fig. 3), then bends toward the south across the "ridge zone" as far as the vicinity of San Carlos de Bariloche. This structure system is composed of fractures developed in the Mesozoic which have caused paleogeographic discontinuities.
2. Across the southern culmination of the megafracture displacing the direction of the Central Valley of Chile the aforementioned system abuts against this and together they turn toward the Pacific joining with the transform fractures of the Chile Ridge (Fig. 3).
3. The double inflection of the unified system corresponds to the contraposition of a homogeneous mass (the north Patagonian massif) with compressive activity around its perimeter and internal tension effects which controlled Jurassic and Cenozoic regional effusive activity.
4. The structural disruption is purely post-Paleozoic and marks the clear and particular differ-

ences in the Patagonian region located to the
south of this transcontinental structure.

With regard to the local geology it is established that:
1. The transcontinental structure described above
marks the southern termination of the geological
provinces situated on its northern flank - the
Southern Sierras of Buenos Aires, the Chaco-Pam-
pean plain, the Pampean Sierras, the Principal
Cordillera, the Central Valley of Chile and the
Coastal Cordillera.
2. The transcontinental structure contains within itself the northern Patagonia coastal area
which acts as a barrier to possible extension of
the former regions and at the same time delimits
the northern extremities of two Patagonian provinces; the North Patagonian and the Patagonian
Cordillera.

These two observations make it possible to suggest that one of the principal factors in the
differentiation between the provinces was the development during the Mesozoic of the structure
transverse to the continent.

The transcontinental structure discussed here
is not the first to be recognized in the South
American continent. Various authors have already
observed the presence of great structural lineaments with east-west trends, especially those related to the origin of the Amazon Basin. In the
last few years these structures have been correlated with similar lineaments in the bottoms of
the Atlantic and Pacific Oceans. In Figure 3,
the positions of these transcontinental structures have been sketched in, in addition to those
described in the present work. To the north of
parallel 22°S the reconstruction of DeLoczy (1977)
is reproduced.

In this figure the ridge system of the eastern
Pacific is also marked together with the whole
area known as the Nazca Plate in the new scheme
of global tectonics.

The coincidence of the location of seismic epicenters with the tension fractures of the ridges
as well as the associated transform faults and
the fractured compressional front at the Pacific
Coast of the continent is now well known.

C. Tectonic analysis of the San Juan seismic zone

For the first time in Argentina a tectonic study
has been made in a seismic area which attempts to
explain the origin and nature of the forces which
generate the earthquakes in the classic mobile
zone of San Juan.

This study (Baldis et al, 1979) has established
the origin of the crustal mobility at medium and
shallow depths which produce the earthquakes in
the area of the Valley of San Juan. It is due to
the opposition of forces in two directions; one
directed from the Pampean nucleus, which is the
principal one, the other from the Andean Arc, displayed in the eastern structure of the Precordillera (forces of reaction). Fig. 3 shows the ar-
rangement of the principal structural elements

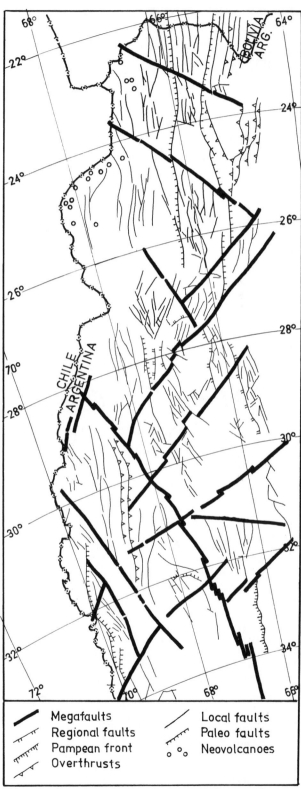

Fig. 2. Geotectonic map between 22° to 34° S

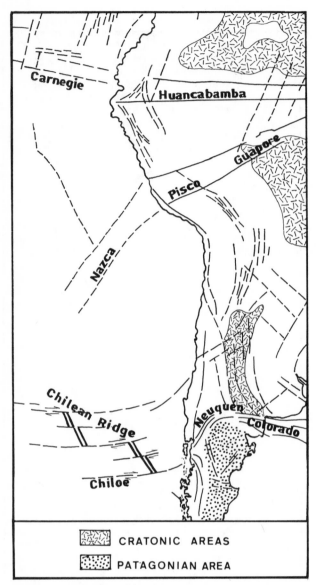

Fig. 3. Position of transcontinental Megafractures.

and the directions of the forces indicating a collision occurs in the neighborhood of the Valley of San Juan (Fig. 1), producing in the Sierra Chica de Zonda a zone of shearing with consequent horizontal components and displacements of blocks. The causes of the permanent seismic activity of the area may be due as much to a local process of accommodation of the Pampean massif as to adjustment to effects produced at the rear of the subduction zone.

Part Two: Advances in Geophysical Knowledge

A. Transcontinental gravimetric profiles at parallels 32° and 36°S.

Following the recommendations on Gravimetry and Earth Tides made at the Conference on Problems of the Solid Earth (Buenos Aires, 1970), the Faculty of Science and Engineering of the National University of Posario prepared transcontinental gravity profiles from the Pacific to the Atlantic.

The first profile was approximately along latitude 32°S (Introcaso and Huerta, 1972). The coverage consisted of:

6 Stations (Chile), University of Chile.
124 Stations (Arg.), National Univ. of Rosario.
188 Stations (Arg.), Instituto Geografico Militar.
 62 Stations (Ura.), Servicio Geografico Militar.
380 Stations, Total.

The total length of the line of profile was 2200 km, the average distance between stations being 6 km.

Along the whole length of the profile a density of 2.67 g/cm^3 was adopted for the Bouguer correction to maintain uniformity since regional effects were sought.

The gravimetric results show small Bouguer anomalies from the center of the profile to the Atlantic Ocean in contrast to the strong negative anomalies which, starting in the Sierras de Cordoba, decrease to an absolute minimum of -295 mgal under the Cordillera of the Andes.

With the object of estimating the thickness of the crust, which is taken to be all the solid material above the Mohorovicic discontinuity, different models were tried to fit the experimental values. The differences between the parameters of these models informs us of the precision obtained in the determination.

From the 62°W meridian to the Atlantic, the thickness of the continental crust is considered to be normal in view of the relatively weak anomalies observed there.

To summarize the interpretations, we have:
Maximum depth of crust: 67±8 km.
Normal thickness of crust adopted: 32±1 km.
Probable density contrast between crust and mantle: 0.39±,04 g/cm^3.

The maximum crustal thickness is found near meridian 70°W in the profile and practically coincides with the axis of the Andean Cordillera.

The second transcontinental gravity profile (Introcaso, 1976) was made along latitude 36°S. In this line values for the Argentine sector are from Instituto Geografico, 1970, and for the Chilean sector from Draguicevic et al, 1961. Simple Bouguer anomalies were used. For the mass correction a value of 2.67 g/cm^3 was employed along the whole profile.

The profile contains points up to 2200 m in altitude along the Argentina-Chile border with corresponding Bouguer anomalies which reach -175 mgals.

The choice of continental crust thickness (normal, or undisturbed) has been hypothetical, using values considered globally acceptable since insufficient seismic data are available.

Adopting densities of 2.9 and 3.3 g/cm^3 for the crust and uppermantle respectively and a normal thickness of 32 km for the crust, the interpretation of the gravity data along latitude 36°S would suggest a maximum crustal thickness of some 50 km. This occurs under the axis of the cordillera, coinciding with the maximum absolute value of gravity recorded. The method of calculation proposed by Introcaso and Huerta (1976) was used in the interpretation.

B. Study of electrical conductivity in the Andes.

The seismic studies carried out by the Carnegie Institution of Washington (CIW) since 1957 showed that seismic waves in passing through the base of the Andean Cordillera undergo severe attenuation. As a working hypothesis (CIW, 1958) it was suggested that this attenuation could be related to the volcanic structure of the Andes and the possible existence of rocks at high temperatures or in a state of partial fusion under the cordillera at relatively small depth.

Within the line of diverse geophysical methods traditionally practiced by CIW. that institution favored studies of electromagnetic induction as a means of seeking a relationship between the anomalous attenuation and a possible anomaly in electrical conductivity. In 1963 the CIW, the Geoohysical Institute of Peru and the Bolivian Geophysical Institute carried out a first differential geomagnetic study with the installation of nine Askania variographic magnetometers (U. Schmucker et al, 1964). A second study was carried out by the same authors in 1965.

Altogether 17 stations were set up in the area bounded by 4°S and 18°S over a strip 500 km. wide parallel to the Pacific coast.

The results (Schmucker et al, 1966; Casaverde et al, 1968) were highly spectular. The nocturnal geomagnetic embayments recorded show an inversion of the vertical component with change of sign of its amplitude. This observed anomaly is opposed to and overcompensates for the effect of the sea coast which would be expected. The axis of the anomaly follows aproximately the axis of the Andean Cordillera and marks the line of concentration of induced currents. The interpretation proposed is of a conducting structure in the form of a cylinder with its axis following that of the anomaly and the cordillera which in this area is approximately NW-SE. Quantitatively, the conductivity proposed for this conducting body is 0.1 ohm $^{-1}$ m $^{-1}$ whilst its top could be up to 50 km in depth.

The agreement between the seismic and conductivity results, both indicating an abnormal thermal state, was undoubtedly a great inducement to extend the study to other sectors of the Cordillera.

In 1972 the CIW, the El Zonda Institute of the National University of San Juan, Argentina, and the Department of Geophysics of the University of Chile installed a simultaneous array of nine magnetometers between latitude 30°S and 33°S, the Pacific coast in Chile and meridian 69°W in Argentina.

The operating procedures were similar to that

of the previous studies based on recording of magnetic embayments and magnetic storms (Aldrich et al., 1972).

The situation found here is different to that encountered previously in the south of Peru and Bolivia (Aldrich et al., 1978). The inversion of sign in the embayments is not found here. The calculation of transfer functions across the three E-W profiles shows a similar tendency in all three characterised by a progressive diminution towards the east in the transfer function of the vertical component with respect to the horizontal components. A regular tendency is also observed in that larger values of the transfer functions are found corresponding to shorter periods in the spectrum.

The induction effect observed, which could originate in deep structures does not show the capacity to overcome the general tendency due to the coast, that is to say, to the sea water as a conductor.

The conditions for electrical conductivity found between parallels 30°S and 33°S, considering the difference from those found in the work done between 4°S and 18°S adds further important evidence about the structural differentiation in the subduction process of the Nazca Plate as it turns back to the N-S direction. Evidence for similar discontinuities is provided by recent seismological studies (Barazangi and Isacks, 1976).

C. The Asthenosphere; Magnetotelluric and Seismic Studies on the Margin of the Subducting Plate.

All modern geodynamic theories concur in postulating an intermediate zone - the asthenosphere - which plays the role in decoupling the lithosphere dynamically from the lower part of the mantle. This permits displacements, both horizontal and vertical, of the lithosphere and the crust.

Many workers at present hold that the asthenosphere, being globally present in the planet, reflects by its horizontal variations in depth, the regional geodynamic conditions. The mechanism of this relation, however, remains in good measure to be clarified.

Within the Plate Tectonic Theory, we will consider a subducting plate margin such as the case of the Nazca Plate beneath South America. Study of the distribution of earthquake hypocenters permits determination of the thickness of lithospheric layers. It has to be admitted the earthquake foci are located in the parts with more rigid properties, such as are attributed to the lithosphere. The rest of the total volume associated with subduction and not occupied by lithosphere could be occupied by material with greater capacity for flowage and incapable of supporting stress to the limit of rupture.

Furthermore, in order to find with precision the limits of the asthenosphere it would be desirable to recognize others of its physical properties. This becomes unavoidable when we move away

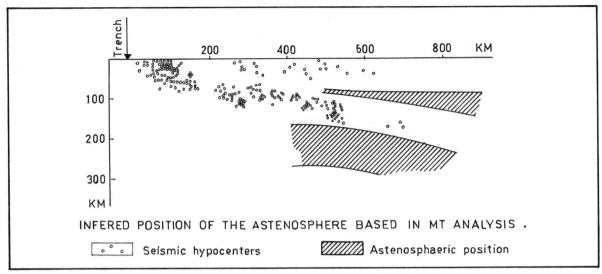

INFERED POSITION OF THE ASTENOSPHERE BASED IN MT ANALYSIS .

Seismic hypocenters   Astenosphaeric position

Fig. 4. Astenosphere and seismic hypocenters based on M.T. profilage.

from the margin of subduction and into the zones considered to be a-seismic which includes a considerable area of the South American continent.

Studies of Rayleigh wave dispersion and regional anomalies in arrival times and seismic refraction studies have established the existence of low velocity layers (LVL) as well as layers of greater attenuation of elastic waves (low Q) in different parts of the world. It is also known that this phenomenon is more intense in geodynamically active areas and would not be present in the more stable or shield areas.

Another tool available for study of the mantle is the measurement of electrical resistivity. The methods of vertical electric logging (VES) with lines of great length and of electromagnetic induction permit the estimation of the distribution of electrical resistivity in the first hundreds of kilometers of depth. Of the electromagnetic induction methods those most used are MT (magneto-telluric) and the group of methods of determining magnetic variations. Both are based on the determination of transfer functions which relate the components of the earth's electromagnetic field to one another. These fields are generated by electric currents in the ionosphere and the magnetosphere.

The MT method is the easiest to operate since measurement stations are occupied individually and successively. The lines are usually 200 meters in length and remain fixed during the course of measurement.

MT measurements made in various parts of the world have resulted in demonstrating the existence of zones of low resistivity (Low Resistivity Layer, LRL) within the upper mantle.

In 1963, with the results at that time available, Fournier et al., (1963) suggested comparision of the LRL with the LVL proposed by Gutenberg (1950).

Later studies have shown that the geodynamic characteristics associated with the LRL - intensification of its effects in active zones, probable absence in shield areas, relationship of the depth of its upper surface to regional heat flow - correspond well to those associated with the LVL, thus displaying a significant comparison between the two phenomena.

In those areas where both magnetotelluric and seismic studies have been studied, it has been found the depths of the LRL and LVL are similar (Fournier, Febrer, 1974). It has also been found that the methods are complementary to each other: the MT method is more effective in determining the upper surface of the LRL whereas the seismic method provides better determination of the base of the LVL.

The international project ELAS (for electrical coductivity of the asthenosphere) which is being developed within Group 3, Division 1 of the International Association of Geomagnetism and Aeronomy (IAGA), coordinates the efforts being made to singularise the asthenosphere by its electrical properties. In turn correlation with seismic and heat flow results is being sought.

D. MT profile in South America across a zone between latitudes 30°S and 32°S.

This section deals with the first profile of MT measurements which is being completed in the zone of destructive contact between two principal lithospheric plates. This is made possible by having access for the study in a continental area, which constitutes a characteristic of South America unique in the world (Figs. 4 and 5).

The profile was begun by the completion of MT measurements in Pilar, Cordoba (Febrer et al., 1977) and later in Cañada de Gomez, Santa Fe and

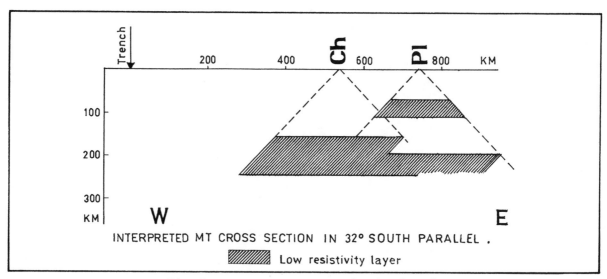

Fig. 5. Low resistivity distribution in 32° S.

Chamical La Rioja (Febrer et al., 1979).

In Figure 5 is shown a transverse section of the zone 30-32°S; on the horizontal axis is marked the distance eastward from the axis of the Chilean trench. The cones drawn in the sketch, whose generating angle is 45°, delimit the effective lateral extension of the measurements. The layers of low resistivity are shaded. In Figure 4 the same transverse section shows the positions of earthquake hypocenters according to Barazangi and Isacks (1976) for the zone 27-33°S. This distribution has the peculiarity that the Benioff zone shows an angle of only 10°.

Let us now test the hypothesis that the low resistivity layers correspond to the volume associated with the asthenosphere.

Subduction of one plate below another, such as we are studying, should show in a transverse section the following elements: lithosphere of South America - the remainder of the asthenosphere of South America - lithosphere of the Nazca plate-asthenosphere of the Nazca plate.

At the base of the layers of low resistivity which have been determined (Fig. 4) what we may call the "electrical asthenosphere" has been drawn in Fig. 5. This gives rise to the thought that the materials which constitute the asthenosphere at the temperature ruling at that depth would have low electric resistance. This suggestion it supported by the results of laboratory experiments using appropriate pressures and temperatures.

Inspection of Figure 5 shows that shaded the zone which we have called the "electrical asthenosphere" does not contain earthquake hypocenters.

The residual asthenosphere or that between the two plates disappears at this latitude by wedging out of the plates, according to the MT results, some 700 Km. from the axis of the Pacific trench. In the model proposed by Isacks et al., 1968, for the Tonga arc the wedging occurs at depth below the line of volcanoes.

In the Andean Cordillera the line of volcanoes is found between 200 to 300 Km. from the axis of the Pacific trench. But precisely between 27 and 33°S there is no Late Tertiary - Quaternary volcanism in the Andes which coincides, as Barazangi and Isacks (1976) point out, with the low angle (only 10°) of the Benioff zone in that latitude.

Conclusion

The MT study in the zone 30-32°S allows us to estimate with the direct evidence of electrical characteristics, the position of the asthenosphere and the thickness of the lithosphere of the Nazca and South America plates at this latitude.

The distribution of earthquake hypocenters (Barazangi and Isacks, 1976) coincides satisfactorily with the MT results presented (Febrer et al., 1979).

References

Aceñolaza, F.G. and A. Bernasconi, Precordillera de Jague, La Rioja, Argentina. Su importancia geologica y estructural, Acta Geol. Lilloana, 11, Nr. 14, 259-290, 1 Abb., 1 Kt., 5 Taf., 3 Prof., Tucuman, 1971.

Aldrich, L.T., J. Bannister, M. Casaverde y E. Triep, Anomalous electrical conductivity structure in Central Chile, CIW Yearbook 1977-8, 1978.

Aldrich, L.T., M. Casaverde, R. Salgueiro, J. Bannister, F. Volponi, S. Del Pozo, L. Tamayo, L. Beach, D. Rubin, R. Quiroga and E. Treip, Electric Conductivity Studies in the Andean Cordillera, Carnegie Institution of Washington Yearbook 1972, 1972.

Angelelli, V., and J.C. Fernandez Lima, Ciclos metalogeneticos de la Argentina, in Geologia Regional Argentina, A.F.Leanza, ed., Acad. Nat. Cienc., Cordoba, Argentina, p. 797-813, 1972.

Arganaraz, R.A., and L.Salazar, Sobre el hallazgo de serpentinitas en la Puna Argentina, Actas V Cong. Geol. Argent. v.l, Villa Carlos Paz, Cordoba, p. 23-32, 1972.

Aubouin, J., and A.V. Borrello, Regard sur la geologie de la Cordillere des Andes; relais paleogeographiques et cycles orogeniques superposes: le Nord argentin, Bull. Geol. Soc. France 7th Series, v. 12, no. 2, p. 246-270, 1970.

Baldis, B., J. Demicheli, J. Febrer, H. Fournier, H. Garcia, E. Huerta, A. Intorcaso and E. Treip, Bilan provisoire pour une synthese geostructural dabs la bande du 32eme parallele a travers l'Amerique du Sud: Geologie structurale, sismologie, gravimetrie, sondage geomagnetique, sondage magnetotellurique, V Reunion des Sciences de la Terre, Rennes, France, 1977.

Baldis, B.A., A. Gorrono, R. Sarudiansky and V. Ploszkiewick, Geotectonica de la Cordillera Oriental, Sierras Subandinas y regiones adyacentes, Actes VI Con. Geol. Arg. 1, 1976

Baldis, B.A., J. Saltify and J.G. Viramonte, Geotectonica de la comarca comprendida entre el Cratogeno Central Argention y el borde austral de la Puna, Actas II Con. Ibero-Amer. Geol. Econ., IV, 1976.

Baldis, B.A., E. Uliarte and A. Vaca, Estructura de la comarca sismica de San Juan, Rev. Asoc. Geol. Arg. XXXIV, (in press)

Barazangi, M., and B. Isacks, Spatial distribution of earthquakes and subduction of the Nazca Plate beneath South America, Geology, V. 4, p. 686-692, 1976.

Barazangi, M., and B. Isacks, Spatial distribution of earthquakes and subduction of the Nazca plate beneath South America, Geology, v.4, p. 686-692, 1976.

Cabre, R., Research in seismology and infrasonic acoustic waves, Final Report, AFOSR, Publ. 26, Obs. San Calixto, La Paz, Bolivia, 87p., 1974.

Caminos, R., Perfil geologico de la cordillera entre 10s 28°00' de latitud sur, provincia de La Rioja, Republica Argentina, Rev. Assoc. Geol. Arg. 27, no. 1, p. 71-83, 1972a.

Caminos, R., Cordillera Frontal, in: Geología Regional Argentina, A.F. Leanza, ed., Acad. Nat. Cienc. Cordoba, Arg., p. 305-343, 1972b.

Carnegie Institution of Washington, Seismic Studies in the Andes, Amer. Geophys. Un. Trans. v. 39, p. 580-582, 1958.

Casaverde, M., A. Giesecke, R. Salgueiro, S. Del Pozo, L. Tamayo and L.T. Aldrich, Anomalia de conductividad bajo los Andes, Geofisica Inter., v. 8, p. 2-4, 1968.

Charrier, R., J. Davidson, A. Mapodozis, C. Palacios, S. Rivano and J.F.C. Vicente, La orogenesis subhercinica: Fase mayor de la evolución paleogeographica y estructural de los Andes Centrales Argentino-Chillenos, V Cong. Geol. Arg., Assoc. Geol. Arg., Resum. S.39, Villa Carlos Paz, Cordoba, 1972.

Chong Diaz, G., and J.J. Frutos, Tectonica y estratigrafia de la Cordillera de Domeyko, Chile, como modelo de la evolucion geologica andina,

V Cong. Geol. Arg., Asso. Geol. Arg., Resum S. 39, Villa Carlos Paz, Cordoba, 1972.

Cingolani, C.A., and R. Varela, Investigaciones geologicas y geochronologicas en el extremo sur de la Isla Gran Malvina, sector del Cabo Belgrano (Cape Meridith), Islas Malvinas, Republica Argentina, VI Cong. Geol. Arg., Resum. 22-23, Bahia Blanca, 1975.

Clark, A.H., J.C. Caelles, E. Farrar, S. McBride and R.C.R. Robertson, Geochronological relationships of mineralization in the central Andean Tin Belt of Bolivia and Argentina, II Cong. Latino-Amer. Geol. Resum., 16-17, Caracas, 1973.

Coira, B., and N. Pezzutti, Vulcanismo cenozoico en el ámbito de la Puna Catamarqueña, Rev. Assoc. Geol. Arg. XXXI, 1, p. 33-52, 1976.

Crowell, J., and L.A. Frakes, Late Paleozoic glacial facies and the origin of the South Pacific Ocean, XXIII Int. Geol. Cong. 13, p. 291-302, Prague, 1968.

Cuerda, A.J., Sierras Pampeanas; una nueva interpretación de su estructura, Rev. Assoc. Geol. Arg., v. 28, no. 3, p. 293-303, 1973.

Demicheli, J., and J. Febrer, Resistivimetro transistorizado para prospeccion geoelectrica, Geoacta, v. 7, no. 2, 1975.

Draguicevic, M. E. Kausel, C. Lomnitz, H. Meinhardus and L. Silva, Levantamiento gravimétrico de Chile, Universidad de Chile, 1961.

Febrer, J., B.A. Baldis, and H. Fournier, Sondaje magnetotelúrico en Chamical, La Rioja, X Reunión Assoc. Arg. de Geofís. y Geodet., San Juan, April, 1979.

Febrer, J., J. Demicheli, and C. Esponda, Acerca de la marcha del proyecto magnetotelúrico del Observatorio Nacional de Física Cósmica, Geoacta, v. 6, no. 3, 1974.

Febrer, J. J. Demicheli, E. García, and H. Fournier, Magnetotelluric sounding in Pilar, Cordoba, Argentina, Acta Geophysica Montanistica, Acad. Sci. Hung., v. 12 (1-3), 1977.

Ferriero, V.J., and R. Mon, Geomorfología y tectónica del Valle de Santa María, Acta Geol. Lilloana, Tucuman, 12, no. 5, p. 73-88, 1973.

Fournier, H., Y Benderitter and J. Febrer, Prospection magnetotellurique dans le massif de Mortagne sur Sevre, Compt. Rend. Acad. Sci. de Paris, Series B, v. 280, p. 141, 1975.

Fournier, H., Y. Benderitter and J. Febrer, Comparaison des mesures de resistivité obtenues par les methodes de sondage electrique et magnetotelluriques en des sites communs situés sur terrain sedimentarie et sur des massif de granite, Mem. du Bur. Res. Geol. et Min., no. 31, 1978.

Fournier, H., and J. Febrer, Magnetotelluric dispersion of results working in log-space, Phys. Earth and Plan. Int. v. 12, p. 359, 1976.

Fournier, H., and J. Febrer, Sondagio magnetotellurico profondo all'Osservatorio Geomagnetico dell'Ebro a Roquetas (Tortosa-Spagna), Atti e Memorie dell'Accademia Patavina de Scienze, v. LXXXIX, Parte II, 1977.

Fournier, H., and J. Febrer, Apport des methodes

d'exploration et d'etude geoelectromagnetique a la connaisance des plaques lithosferiques: leur contenu, leur substratum et leurs zones de contact, VI Reunion Sci. de la Terre, Paris, 1978.

Fournier, H., and J. Febrer, Sondaje magnetotelúrico profundo en el Observatorio geomagnético de Toledo (Espana), Vol. Com. 2a, Assemb. Nac. de Geodesia y Geofísica, Madrid, 1978.

Frutos, J., Evolución y superimposition tectónica en los Andes Chileno-Argentino-Bolivianos, II Cong. Latino-Amer. Geol. Resum. 62-63, Caracas, 1973.

González Díaz, E., Aspectos geológicos del límite argentino-chileno entre los paralelos 27°00' y 27°30' de latitud sur (SO de la Provincia de Catamarca), VI Cong. Geol. Arg. Resum. S.31, Bahia Blanca, 1975.

Instituto Geográfico Militar Argintino, Actividades gravimétricas (1956-1970), 1970.

Introcaso, E., Modelo gravimétrico provisorio de corteza andina en el paralelo 36°S, Geoacta, v. 8, no. 1, 1976.

Introcaso, A., and E. Huerta, Perfil gravimétrico transcontinental sudamericano, Revista Cartográfica del PAIGH, no. 22, Univ. Nac. de Rosario, 1972.

Introcaso, A., and E. Huerta, Valuación de effectos gravimétricos y aplicaciones a la interpretación, Geoacta, v. 8, no. 1, 1976.

Marturet, R., and J. Fernández, Problemas tectónicos de la Puna Oriental I; El sinclinal de Mal Paso y la Sierra de Aguilar, Actas V Cong. Geol. Arg. v. 5, p. 57-69, Buenos Aires, 1973.

Mendez, V., Estructuras de las Provincias de Salta y Jujuy a partir del meridiano 65°30' Oeste, hasta el límite con las Repúblicas de Bolivia y Chile, Rev. Assoc. Geol. Arg. 29, no. 4, p. 391-424, Buenos Aires, 1974.

Mendez, V., A Navarini, D. Plaza and V. Viera, Faja eruptiva de la Puna Oriental, Actas V Cong. Geol. Arg., 4, p. 89-100, Buenos Aires, 1973.

Mon, R., Estructura geológica del extremo austral de la Sierras Subandinas, Provincias de Salta y Tucumán, República Argentina, Rev. Assoc. Geol. Arg. 26, no.2, p. 209-220, Buenos Aires, 1971.

Mon, R., Estructuras de las Sierras Subandinas Argentinas, II Cing. Latino-Amer. Geol., Resum., 37-38, Caracas, 1973.

Mon, R., Esquema tectonico de los Andes del Norte Argentino, Rev. Assoc. Geol. Arg. XXXIV, 1, p. 53-60, 1979.

Mon, R., and A. Ardaneta, Geología del borde oriental de los Andes en las Porvincias de Tucumán y Santiago del Estero, República Argentina, II Cong. Latino-Amer. Geol. Resum. 36-37, Caracas, 1973.

Nemec, V., Regular structural patterns in Argentina, VI Cong. Geol. Arg. Resum., S.41, Bahia Blanca, 1975.

Palmer, K.F., and I.W.D. Dalziel, Structural studies in the Scotia Arc: Cordillera Darwin, Tierra del Fuego, Antarc. Jour., v. 8, no. 1, p. 11-14, Washington, 1973.

Parada M, and H. Moreno, Geología de la Cordillera de los Andes entre los paralelos 39°30' y 41°30'S, VI Cong. Geol. Arg. Resum., s.33, Bahia Blanca, 1975.

Pasotti, P., Neotectónica en el llanura pampeana, II Cong. Latino-Amer. Geol. Resum., 21, Caracas, 1973.

Ramos, V.A., Estructura de los primeros contrafuertes de la puna Salto-Jujeña u sus manistaciones volcánicas asociadas, Actas V Cong. Geol. Arg., 4, p. 159-202, Buenos Aires, 1973.

Ramos, V.A., El volcanismo del Cretácico inferior de la Cordillera Patagónia, Actas VII Cong. Geol. Arg., I, p. 423-436, 1979.

Schmucker, U., S.E. Forbush, O. Hartmann, A. Geisecke, M. Casaverde, J. Castillo, R. Salguero and S. Del Pozo, Electrical conductivity under the Andes, Carnegie Inst. Wash. Yearbook, 65, p. 11-28, 1966.

Schmucker, U., O. Hartmenn, A. Geisecke, M. Casaverde and S. Forbush, Electrical conductivity anomalies in the Earth in Peru, Carnegie Inst. Wash. Yearbook, 63, p. 354-362, 1964.

Schwab, K., Die Stratigraphie in der Umgebung des Salar de Cauchari (NW Argentinien), Geotekt. Forsch. 43, no. 1-2, p. 1-168, Stuttgart, 1973.

Turner, J.C.M., and B.A. Baldis, La estructura transcontinental del norte de Patagonia, Actas VII Cong. Geol. Arg., II, 1979.

Varela, R., Estudio geotectónico del extremo de la Precordillera de Mendoza, República Argentina, Rev. Assoc. Geol. Arg., XXVIII, no. 3, p. 241-267, Buenos Aires, 1973.

Varela, R., Sierras Australes de la Provincia de Buenos Aires: Hipótesis de trabajo sobre su composicion geológica y rasgos tectónicos salientes, Rev. Assoc. Geol. Arg. XXXIII, 1, p. 52-62, 1978.

Vicente, J.C., Observaciones sobre le tectónica de la Cordillera Principal de San Juan de Mendoza entre el Cerro Mercedario y el Río Blanco del Tupungato (32° a 33° latitud sur), V Cong. Geol. Arg., Assoc. Geol. Arg. Resum., S.42, Villa Carlos Paz (Cordoba), 1972.

Vicente, J.C., L'orogene Hercynien de l'Amérique meridionale: Essai sur l'organisation paléogéographique et structurale des Cuyanides, II Cong. Latino-Amer. Geol. Resum., 3-5, Caracas, 1973.

Vicente, J.C., R. Charrier, J. Davidson, A, Mapodozis ans S. Rivano, La orogenesis Subhercinica: fase mayor de la evolucion paleogeografica y estructural de los Andes Argentino-Chilenos Central, Actas V Cong. Geol. Arg., 5, p. 81-98, Buenos Aires, 1973.

PROGRESS IN GEODYNAMICS IN THE SCOTIA ARC REGION

Peter F. Barker

Department of Geological Sciences, Birmingham University,
Birmingham B15 2TT, England

Ian W.D. Dalziel

Lamont-Doherty Geological Observatory, Palisades,
New York   10964 U.S.A.

Abstract. Considerable progress was made over
the period of the Geodynamics Project, both in
understanding the tectonic evolution of the
Scotia Arc region and in studying those active
margin processes which are particularly clearly
developed there. The region divides naturally
into two independent geological provinces. The
Pacific margin province, including almost all of
southern South America and the Antarctic
Peninsula, began its development at the Pacific
margin of Gondwanaland perhaps as early as the
Palaeozoic and has remained in that environment,
of probably episodic subduction of Pacific
oceanic lithosphere, ever since. The Scotia
Sea province does include elements of that
original Pacific margin province, as inert continental fragments within the Scotia Ridge, but
has developed essentially independently as a
complication on the South American-Antarctic
plate boundary over the past 40 Ma. This report
first describes what is known of the evolution
of these provinces, to set the scene for an
account of those active margin processes which
have been studied extensively, and to draw
attention to other examples as yet unexamined
in detail.

The Pacific margin province contains well-
developed examples of ensialic back-arc basins,
in the Mesozoic "Rocas Verdes" ophiolite complex
of southern Chile and the currently opening
Bransfield Strait, which have stimulated interest
in ophiolite geochemistry, ocean-floor structure
and metamorphism. The young intra-oceanic South
Sandwich island arc and back-arc basin have also
been studied geochemically and the eventual
closure of the Rocas Verdes basin provides an
interesting range of examples of Cordilleran
deformation. The volcanic and plutonic rocks of
the Antarctic Peninsula have been used to
fomulate a geochemical model of magma genesis
above subduction zones.

A. Introduction

The particular concern of Working Group 2 of
the Geodynamics Project has been the study of
dynamic processes at the western margin of the
Americas, continents with a long history of
episodic subduction of Pacific oceanic lithosphere.
The Scotia Arc region is a natural sub-division,
characterised by a wide variety of active margin
processes, including some which are young and
therefore relatively simple. The region has other,
less desirable characteristics, notably a harsh
climate and a logistic inaccessibility which
restrict and distort field studies. These
difficulties have also dictated the general
working style of the Scotia Arc Study Group,
so that the contribution of the Geodynamics Project in the Scotia Arc region has lain less in
concrete projects jointly conceived and carried
out than in bending the thoughts of people
working there away from mere mapping and classification, towards the understanding of geodynamic
processes.

The term "Scotia Arc" has been applied to all
or part of the long eastward loop of islands
and submarine ridges which surrounds the Scotia
Sea on three sides. The much broader definition
of its proper domain assumed originally by the
Study Group (and followed in this report) included the South American and Antarctic
Peninsular margins between 50°S and about 75°S,
all of the Scotia Sea and the surrounding Scotia
Ridge (only a small part of which is an island
arc in the strictest sense). This area appeared
to contain a wide variety of plate margin interactions, past and present, which would prove
amenable to more precise definition by complementary onshore and offshore investigations.
The more successful of these are considered in
detail in the later parts of this report, but we
first need to describe what is known of the

geological evolution and present tectonic state of the region, to provide a general context for the more specialised studies. That context is itself largely the product of the years of the Geodynamics Project, and its description here also provides the chance to note where the region's potential for the study of processes remains unexploited.

The region divides naturally into two provinces whose tectonic evolution has been distinctly different. The land geology of southern South America and the Antarctic Peninsula results from their original location at the Pacific margin of Gondwanaland, before its fragmentation, and the continuing but probably episodic subduction of Pacific oceanic lithosphere ever since. The Scotia Ridge and Scotia Sea, on the other hand, exist at present as a number of small plates which complicate the boundary between the much larger South American and Antarctic plates, and probably evolved in that same environment over the last 30 or 40 Ma. At present there is no evidence that the two provinces did not develop independently of each other, although many aspects of the earlier history of the region are still obscure.

B. Tectonic setting and geological evolution

B1. Scotia Arc Region – present major plate motions and boundaries

Although much more is known about them now than in 1970, present plate boundaries and motions in the Scotia Sea region are not yet completely understood. The boundary between the South American (SAM) and Antarctic (ANT) plates extends from the Bouvet triple junction (B in Figure 1) in the South Atlantic, by way of the Scotia Arc region, to the junction of the Chile Rise (C) and Peru-Chile trench at about 45°S. As Figure 1 suggests, the boundary is simple only in its most easterly part, between the Bouvet triple junction and the southern end of the South Sandwich trench. Over this length it has the form of a spreading centre with long, east-west dextral offsets of the ridge crest (Barker, 1970; Forsyth, 1975; Sclater et al., 1976), and it is from this length alone that the data have come (in the forms of earthquake first motions, fracture zone orientations and spreading rates) which have been incorporated into the more recent, and better computations of worldwide plate motions. In such computations, the apparent lack of control over the remainder of the boundary is compensated by data from other plate boundaries. The SAM-ANT motions computed by the two most recent models are very similar, and agree with the local data. Chase (1978a) computes a rotation of 0.344 deg/Ma about a pole at 87.1°S, 48.0°W, and Minster and Jordan (1978) obtain values of 0.302 deg/Ma about a pole at 87.7°S, 104.8°W. Computed uncertainties (2σ)

are of the order of 0.05 deg/Ma and 7 degrees, so that the motions are essentially identical.

Over most of the Scotia Sea region these motions amount to an eastward movement of the Antarctic plate with respect to the South American plate. Between about 25° and 70°W, however, the SAM-ANT boundary as such does not exist. The intense earthquake activity associated with subduction at the South Sandwich trench gives way to a small number of earthquakes along both the north and the south Scotia Ridge and an apparently diffuse distribution in western Drake Passage (Figure 1). We show later with the aid of magnetic anomalies, first motion data and land geological information, that these are associated with the boundaries of up to four small plates, the motions of which are partly decoupled from SAM-ANT motion. Two of the small plates lie within the eastern, Scotia Sea province, and the other two are directly adjacent on the western side of the Shackleton fracture zone (S in Figure 1), but are essentially part of the Pacific margin province, which is described next.

B2. Southeast Pacific – spreading and subduction history

The land geology of South America, West Antarctica, New Zealand and Australia reveals

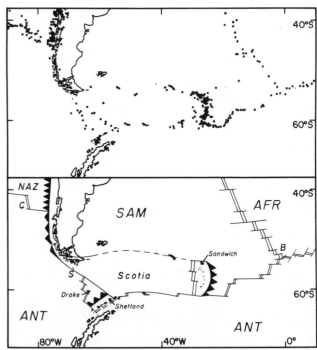

Fig. 1. Shallow earthquakes and plate boundaries in the Scotia Arc Region, compiled mainly from Barker (1970, 1976), Forsyth (1975), Sclater et. al. (1976), Herron et al. (1977), Barker et al. (1982). SAM-ANT motion is sinistral, slow (20-24 mm/year) and approximately east-west.

in a general sense   a history of subduction
at their Pacific margins which extends back into
the Palaeozoic. The ocean floor magnetic
record in the Pacific goes back only as far
as the Jurassic, and because of more recent
subduction is unavoidably incomplete. In the
Scotia Arc region, a more detailed history of
Pacific margin subduction can be obtained from
magnetic anomalies only for the Cenozoic.

Most of the ocean floor of the South Pacific
was formed over the past 80 Ma at the Pacific-
Antarctic ridge (Weissel et al, 1977) but in
the east, adjacent to southern South America
and the Antarctic Peninsula, two different
spreading systems are seen (Figure 2 and
Barker, 1982; Cande et al, 1982). That off
South America appears in all respects to have
been part of a southerly extension of the Chile
Rise (Herron and Hayes, 1969; Herron, 1971).
The associated anomalies have a NNW trend, and
young eastward, indicating that the spreading
centre migrated into the trench. The ages of the
oldest anomalies in the west, and the youngest
anomalies at the trench, both increase southward. Thus, the Chile Rise appears as a
transient phenomenon linking the Pacific-
Antarctic Ridge to the trench, and migrating
northward along the South American margin. Its
passing meant that a fast subduction regime
akin to that now seen north of 46°S (>90 mm/year-
Minster and Jordan, 1978) was replaced to the
south by sections of the SAM-ANT boundary.
Present day SAM-ANT motion, already mentioned,
here becomes east-west convergence, presumably
subduction, but at the much slower rate of 20
to 24 mm/year. A few large, shallow earthquakes are recorded from this part of the margin,
and reflection profiles (Hayes and Ewing, 1970)
show oceanic basement dipping toward the trench,
but overlain by up to 2 km of sediment which
completely obliterates the trench topography.
The slow subduction, which these features suggest,
would have replaced fast subduction at some time
after the age of the oldest magnetic anomaly now
seen at the margin. Barker (1982) estimates
that between 50° and 54°S the ridge crest was
subducted about 13 Ma ago. An improved understanding of the history of SAM-ANT motion will
permit a better estimate, although subduction
south of 52°S will be slower and oblique,
reflecting motion between the Antarctic and much
smaller Scotia plate (Figure 1 and Section B6).
The area near 46°S, where the current ridge
crest-trench encounter is taking place, has been
described by Herron et al (1981).

Off the Antarctic Peninsula lie several
sections of NE-trending magnetic anomalies,
bounded by NW-SE fracture zones (Figure 2).
Sections become generally younger northeastward,
and form two distinct tectonic provinces. In
the more northeasterly province, between the
Hero and Shackleton fracture zones, ocean floor
youngs away from the Peninsula, and both flanks
of the spreading centre are seen (Barker, 1970,

1982; Barker and Burrell, 1977). This part of
the margin includes a topographic trench. In
the southwestern province (Herron and Tucholke,
1976; Cande et al, 1982), magnetic anomalies
young towards the margin, indicating the former
migration of the spreading centre into the
trench. The margin here differs from that off
southern Chile, however, in that reflection
profiles show oceanic basement at the margin
to lie horizontally, rather than to dip into a
trench (Herron and Tucholke, 1976). Moreover,
basement depth at the margin is similar to that
of normal ocean floor of the same age (Barker,
1982), suggesting that subduction simply stopped
as each section of ridge crest reached the
trench.

Magnetic profiles are not abundant in this
southwestern province, and spreading rates
change frequently, but there is sufficient overlap between adjacent sections to demonstrate that
the opposite, subducted flanks of all sections
of the spreading centre were part of a single
plate. This plate (the "Aluk" plate of Herron
and Tucholke, 1976, but probably a continuation
of the Cretaceous mid-Pacific "Phoenix" plate:
Barker, 1982) is now represented at the surface
only as the small "Drake" plate off the South
Shetland Is. (Figure 1).

Times at which sections of spreading centre
arrived at the trench are not known everywhere
directly, but may be estimated using anomalies
from adjacent sections. Arrival times are
progressively younger northeastward, from 50 Ma
near the Tharp fracture zone to only 4 Ma
directly SW of the Hero fracture zone (Figure 2).
The geometry of subduction was unusually simple,
so that subduction rates and the age of ocean
floor being subducted at any time may also be
estimated (at least for the Cenozoic). This
provides opportunities for a quite detailed and
precise comparison between subduction dynamics
and the onshore geology (considered in Sections
B5 and C2). Another such opportunity is
presented by the very young (4 Ma) ridge crest-
trench collision off Brabant I., to test speculation that parts of an older fore-arc were
tectonically eroded and subducted during ridge-
crest approach (Barker, 1982), and to study the
relationship between subduction and mineralisation (Hawkes, 1982).

That length of the Antarctic Peninsula
opposite the younger, northeasterly magnetic province is the only length possessing a topographic trench, onshore late Tertiary volcanism
(on the South Shetland Is.) and a postulated
back-arc extensional basin along Bransfield
Strait. However, this does not reflect a simple
continuation of the earlier tectonic regime,
in which successive sections of the Peninsular
margin became welded to the ocean floor as the
ridge crest reached the trench, while subduction
and spreading continued farther along the margin
to the northeast. In contrast, spreading in
this province also stopped 4 Ma ago (the time

Fig. 2. Ocean floor magnetic anomalies in the southeastern Pacific, from Barker (1982). Preferred interpretations of earthquake focal mechanisms from Forsyth (1975).

of the most recent ridge crest-trench collision), with the three ridge crest sections en echelon along 65°W, some distance from the trench. Earthquakes are associated with the Shackleton fracture zone and with the other ridge crest and intervening transform sections (Figure 1) but it is not obvious that these denote continued (albeit slow) spreading, since the only earthquake sufficiently well-recorded for first motion studies (Forsyth, 1975 and Figure 2) shows WSW-ENE compression and may be intraplate. A maximum of 20 km of ocean floor has been produced at the ridge crest since 4 Ma (Barker, 1976). The 4 Ma collision was the first after the 6 Ma stoppage of spreading in eastern Drake Passage, on the far side of the Shackleton fracture zone (Barker and Burrell, 1977), and other changes farther east in the Scotia Sea (Section B6). Exactly how these changes modified the balance of forces on the small Drake or remnant Aluk plate is uncertain, but the imposition of WNW-ESE compressional stress at the long fracture zone boundaries of the plate could have been decisive in stopping further spreading.

No earthquakes have been detected on the plate dipping beneath the South Shetland Is, but those in Bransfield Strait, together with the volcanism along the axis of the basin, suggest that back-arc extension is still active. Such activity implies continued subduction, and defines a separate small Shetland plate (Figure 1). It may therefore be argued that in this case, as has been proposed elsewhere (Chase, 1978b for example), back-arc extension started in response to a requirement for continued sinking of the subducted slab. The nature of the extension in Bransfield Strait is interesting in view of its youth and restricted extent, and is considered in detail in Section C1b.

Thus, much of the Cenozoic subduction history at the Pacific margin of southern South America and the Antarctic Peninsula is known, and additional survey should refine this information further. It may become possible, from studies of the entire Pacific Basin, to argue that subduction must have been taking place at a particular margin at a particular time in the more distant past. Because of the likelihood of the subduction of entire spreading regimes, however, it will never be possible to deduce that subduction was not taking place, and detailed studies of the relationship between subduction and igneous activity seem likely to be confined to the Cenozoic.

B3. Pacific margin - pre-Jurassic setting

The marine geophysical survey and earthquake studies on which earlier sections of this report have been based were nearly all carried out and described during the period of the Geodynamics Project. This is not so for the land geology; although most of the detailed, process-oriented studies on which this report concentrates are

recent, they rest on an extensive base of reconnaissance and detailed geological mapping and stratigraphic correlation, which cannot be described here but which must be acknowledged.

As yet, there is no generally accepted original position of the Scotia Arc region within Gondwanaland. Different and mostly mutually exclusive reconstructions and models of subsequent evolution have been proposed by Dietz and Holden (1970), Smith and Hallam (1970), Dalziel and Elliott (1971, 1973), Barker and Griffiths (1972, 1977), Barker, Dalziel et al. (1977), de Wit (1977), Barron et al. (1978) and Norton and Sclater (1979). These interpretations are based to a varying and sometimes negligible extent on data from within the region itself, and in the absence of consensus it is not wise to rely upon them alone for insights about the early tectonic environment. A review (Dalziel, 1982) of the pre-M. Jurassic rocks of the Scotia Arc region argues that, although exposures are few and scattered, they do nevertheless give a firm indication of the tectonic environment of the region before Gondwanaland break-up. Dalziel's essential conclusion is that the pre-M. Jurassic 'basement' rocks of southern South America and the Antarctic Peninsula represent portions of the arc-trench gap and main magmatic arc of a late Palaeozoic-early Mesozoic subduction zone along Gondwanaland's Pacific margin. The fore-arc environment is represented by greywackes and shales from a volcano-plutonic terrain of largely granitic composition, chert-argillite and metavolcanic-chert-marble-calcsilicate sequences, mafic pillow lavas and tectonically-emplaced dunite-serpentinite, showing metamorphism up to blueschist facies in places and affected by polyphase deformation. Sedimentary rocks of Carboniferous to Triassic age are involved and exposures stretch from Alexander I through the South Shetland Is to the South Orkney Is and along much of the Chilean margin south of 50°S (Figures 3 and 4). Pre-M. Jurassic plutons occur less widely, but are seen in both South America and the Peninsula, and in almost all cases lie on the 'continental' side of the fore-arc rocks, indicating a polarity for the subduction zone identical to that of the present day.

Since the above-mentioned review was presented, similar interpretations have been published by Smellie (1981) and Hyden and Tanner (1981). Isotopic studies of the South Orkney Is and South Shetland Is metamorphic complexes, however, indicate that parts of the fore-arc terrain may be late Mesozoic or Cenozoic in age (Tanner et al, 1982).

The pre-M. Jurassic provinces in southern South America have also been defined more precisely by Forsythe (1982).

Dalziel (1982) notes that deformation in an arc-trench gap is likely to be continuous while subduction is taking place, but detects a change of some kind in the Late Triassic-Early Jurassic,

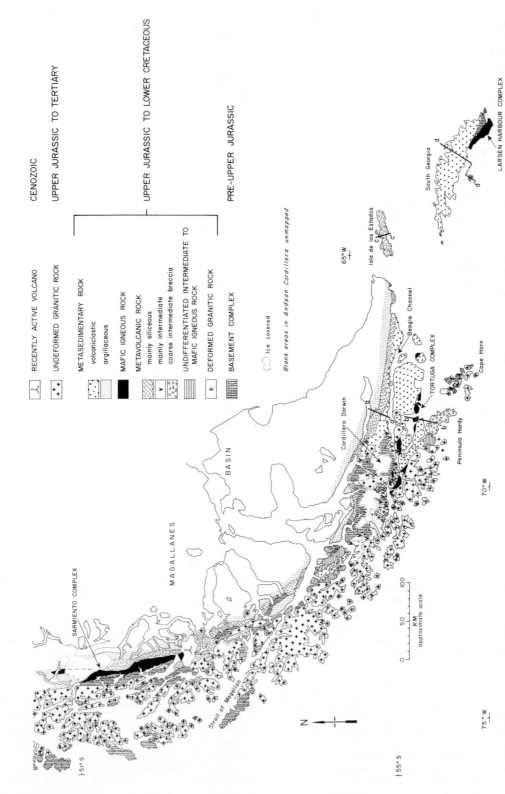

Fig. 3. Geological map of southern South America, with South Georgia in restored position and locations indicated of structural sections in Fig. 12. (Dalziel, 1981).

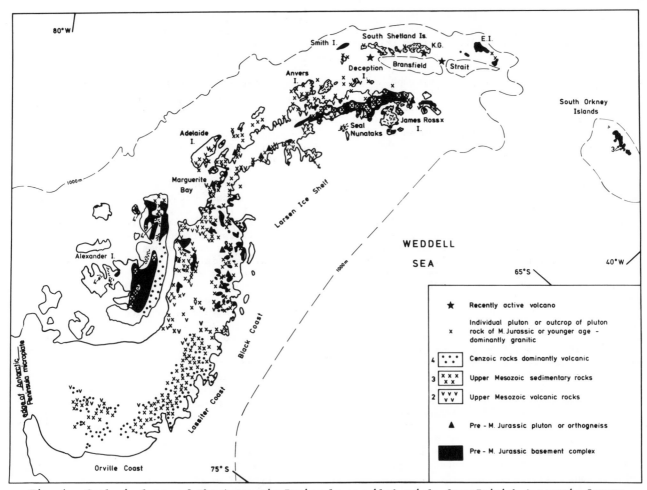

Fig. 4. Geological map of the Antarctic Peninsula compiled mainly from British Antarctic Survey maps. (British Antarctic Survey, 1979, 1981, 1982).

accompanied by uplift and erosion of all of the tectonic elements discussed above. The Middle Jurassic saw renewed and extensive calc–alkaline volcanic and plutonic activity, the locus of which jumped Pacificward from its earlier position, and which was distinctly more silicic away from the Pacific in both the Antarctic Peninsula (Adie 1972) and southern South America (Dalziel, 1974; Bruhn et al., 1978). The onset in the Peninsula (Rex, 1976) may have been 10 to 20 Ma earlier than in South America (Halpern, 1973).

B4. Southern South America : post-Triassic evolution

Subaerial, mainly silicic volcanism covered a zone up to 600 km wide in southern South America during the middle to late Jurassic (Bruhn et al., 1978) and there is much discussion as to whether this exceptional thermal event was essentially subduction-related or an early phase of Gondwanaland break-up. Faulting and subsidence took place directly behind the island arc during

the later stages of this activity, and in the latest Jurassic or earliest Cretaceous there developed a back-arc basin with a partly mafic floor (Dalziel et al., 1974), which is also seen in South Georgia (Storey et al., 1977). The evolution of this basin, and in particular the nature of its ophiolitic floor, the Rocas Verdes Complex, have been studied intensively, and are described in more detail in Section C1c, below. The basin was filled with rapidly-deposited volcaniclastic sediments from the island arc (the Yahgan formation of South America and Cumberland Bay sequence of South Georgia), with an interfingering of a lesser, more siliceous component from the cratonic, Atlantic side (Sandebugten sequence of South Georgia). In Albian-Coniacian times there was simultaneous subsidence of the foreland region, and uplift and (to an extent not yet understood) translation of the Pacific side of the basin towards the foreland. The overlying sediments and adjacent continental basement were intensively deformed (see Section C3) and, with the island arc and ophiolite complex, became sediment sources for the new (foreland) basin (Natland et al., 1974, Dott et al., 1982). By then this basin extended over the present area of the Magallanes and southern Malvinas basins (with South Georgia lying directly east of Cape Horn – Figure 3) and for an unknown distance farther east, on and to the south of the Falkland Plateau (Barker, Dalziel et al., 1976). Sediments from the rising Cordillera were deposited as deep-sea fan turbidites, and there was a considerable component of current flow along the basin axis from the north (Scott, 1966; Dott et al., 1982).

The subsequent history of the Magallanes basin is one of continued but probably episodic uplift of its Pacific side, shown by the involvement of progressively younger material from the Cordillera in the basin sediments, and by the eastward migration of the basin's axis (Natland et al., 1974). The basin generally shallowed from the north over the same period. Calc-alkaline volcanic detritus within these sediments and radiometrically dated rocks from the Patagonian batholith indicate continued subduction-related igneous activity into the Miocene. The deformed rocks were overfolded and thrust towards the Atlantic foreland shelf, with slow Atlanticward migration of the deformational front. The history of deformation is difficult to assess, but Winslow (1982) infers some kind of pause between the mid-Cretaceous uplift and deformation of the marginal basin and a mid-Tertiary (post Eocene) episode. Unlike the earlier episode, the mid-Tertiary deformation did not appear to involve the basement rocks, and a décollement underlying the deformed sediments is implied. Deformational intensity increases along the basin axis towards the southeast, but the relative importance in this of uplift of the Pacific margin and crustal shortening is not yet known.

A more precisely mappable feature of the mid-Tertiary folding and thrusting is the progressive younging, northwestward along the basin axis, of the youngest rocks involved. At the two extremes, Eocene sediments are deformed, Miocene sediments undeformed in southernmost Tierra del Fuego, but at about 52°S even Plio-Pleistocene basaltic dykes and flows are gently folded, Pliocene tuffs strongly folded. Winslow suggests that this diachroneity reflects the northward migration along the margin of the Chile Rise/ Peru-Chile trench triple junction, noted in Section B2.

Superimposed upon the deformation described above (Figure 5) is a younger system of faulting, dominantly strike-slip but also including high-angle reverse faults. There is ample evidence from the land geology and earthquake first motions of active faulting of this kind (Forsyth, 1975; Fuenzalida, 1976; Winslow, 1982) and up to 80 km of sinistral strike-slip displacement may have occurred since the Jurassic along one such fault, that occupying the western

Straits of Magellan. Displacements of 1-8 km are well documented. The precise chronology of the younger fault system is made obscure by its tendency to reactivate faults of the older deformational systems, and the possibility that these older systems also had a strike-slip component in places (Winslow, 1982). The younger system is in fact the western end of the SAM-Scotia plate boundary which runs eastward into the north Scotia Ridge, so that information about its onset would be a valuable aid to understanding Scotia Sea evolution (see Section B6).

The Patagonian batholith forms the southern end of the Andean batholith of South America. It is about 100 km wide and its major rock type is tonalite, comparable in composition to the average circum-Pacific andesite (Stern and Stroup, 1982). It intrudes the pre. M. Jurassic 'basement' and includes bodies both older and younger than the Rocas Verdes marginal basin. It occurs mostly on the Pacific side of the marginal basin, but not entirely. Halpern (1973) reports three main igneous episodes at 155 to 120 mA, 100 to 75 Ma and 50 to 10 Ma, but detects no simple east-west progression in age such as occurs in northern Chile (Farrar et al., 1970; McNutt et al., 1975) or in the Antarctic Peninsula (Rex 1976).

Little detailed work has been done on the Patagonian batholith. No individual plutons have been studied in detail and the only body of petrologic/geochemical data comes from a traverse between 51° and 52°S by Stern and Stroup (1982). They detect an eastward increase in K_2O and Ce/Yb ratio at similar SiO_2 contents, as is typical of circum-Pacific subduction-related batholiths. By excluding those plutons older than marginal basin formation, on the grounds that they may have formed by crustal anatexis in association with the very widespread Late Jurassic acid volcanism, Stern and Stroup are able to see some semi-systematic migration of intrusive activity with time. The youngest plutons occur to the east of the marginal basin, in the Andean foothills but both Cretaceous and Tertiary plutons occur in the area closest to the Pacific margin. The zone of plutonism therefore appears to be widening eastward, rather than simply migrating, and is thus intermediate in character between the eastward-migrating northern Chilean batholith and the westward-migrating Antarctic Peninsula batholith (see Section C2).

One very interesting feature of the Pacific margin of southern Chile is the absence of any exposures on land of material originating in the post-M. Jurassic arc-trench gap. The continental shelf is narrow south of 45°S and the westernmost onshore exposures are of batholithic rocks or of elements of the pre-M. Jurassic arc-trench gap (Bruhn and Dalziel, 1977 and Figure 3). A series of mid-slope terraces along the continental slope off southern Chile (Hayes, 1974) closely

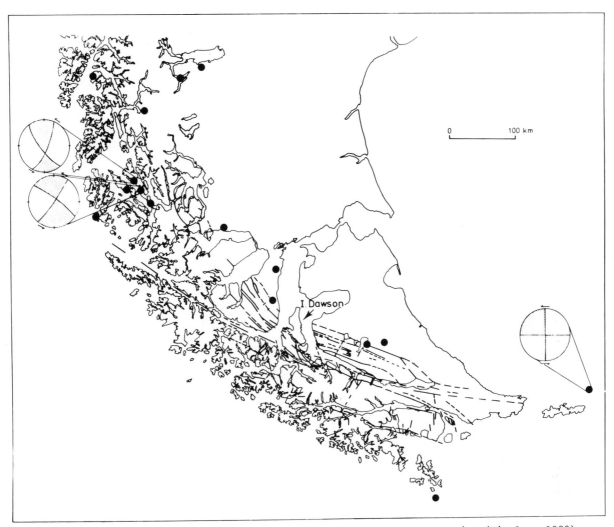

Fig. 5. Cenozoic and recent faulting and seismicity in southern South America (Winslow, 1982).
Solid lines are fault traces with documented recent motion. Small circles are relocated earthquakes;
focal mechanisms are from Forsyth (1975).

resembles those of subducting margins elsewhere, as described, for example, by Karig (1977). This interpretation of their origin would be entirely consistent with the view of present plate boundaries illustrated by Figure 1 and suggested in section B6, in which the SAM-Scotia -ANT triple junction occurs at the western end of the Straits of Magallan, and slow subduction continues at both the SAM-ANT boundary to the north and (obliquely and even more slowly) at the Scotia-ANT boundary to the south. Herron et al. (1977) however, argue that subduction to the south has ceased and the mid-slope terraces are merely the offshore ends of a complex series of repeated sinistral strike-slip faults which are dissecting the Pacific margin of Tierra del Fuego. Under this interpretation, the triple junction occurs inland, near Isla

Dawson (Figure 5). Winslow (1982) makes the interesting comment that present-day sinistral strike-slip faulting in this complex region is accentuating the overall eastward bend of structures in southernmost South America, the age and origin of which have provoked much discussion because of their possible signific-ance for Gondwanaland reconstruction (for example, Dalziel and Elliott, 1971; Barker and Griffiths, 1972; Dalziel et al., 1973; Barker, Dalziel et al, 1976; de Wit, 1977, Dalziel 1982). Barker and Burrell (1977) suggest that the more elevated parts of the Shackleton fracture zone (S in Figure 1) are of continen-tal composition, and were detached from the lower arc-trench gap of the Pacific margin of Tierra del Feugo by strike-slip motion during Scotia Sea opening.

As discussed above, calc-alkaline plutonic
and volcanic activity along the Antarctic
Peninsula may have started somewhat sooner,
after the late Triassic-early Jurassic uplift
and erosion noted by Dalziel (1982), than it
did in southern South America. Plutons
provide most of the exposed rock of the
Peninsula (Figure 4) and are particularly
dominant along its spine. It may be only a
matter of ice cover and erosion level that
they do not have the dimensions of the Andean
batholith. Rex's (1976) important compendium
of K-Ar radiometric ages of Antarctic
Peninsula plutons north of 70°S shows four
groups, 180 to 160 Ma, 140 to 130 Ma, 110 to
90 Ma and 75 to 45 Ma, which do not correspond
entirely to Halpern's (1973) groupings from
southern South America. The plutons are
systematically distributed, the greatest ages
being restricted to the east coast and,
especially in the north, the least ages to the
west coast. Volcanic rocks are widely
distributed also, but often are exposed only
where intruded by younger plutons, so that
they are only of relative stratigraphic
value. These cases aside, ages range from
183 Ma to the present. The ages show the same
geographical distribution as for the plutons
(if less completely and precisely), and the
Mesozoic lavas are systematically more silicic
towards the east as in South America. More
recent Rb-Sr determinations by Pankhurst (1982b)
generally confirm the time- and space-distribu-
tions of the K-Ar data, although Pankhurst
(1982b) emphasizes the essential continuity
of magmatic activity from late Triassic times.
The availability of a large number of dated
samples and the recognition that the ages are
to some extent systematically distributed (and
in the opposite sense to northern Chile -
Farrar et al., 1970) has led to a major geochemical study of Antarctic Peninsula plutonic
and volcanic rocks of considerable geodynamic
interest, which is reported in Section C2
below.

The renewal of calc-alkaline volcanic
activity along the Peninsula in the Jurassic and
its continuation through the Cretaceous would
have led to the production of a considerable
volume of volcaniclastic sediments, but these
are somewhat scantily and unevenly preserved
in the geological record. Volcaniclastic
sediments in the northern part of the Antarctic
Peninsula are thin: upper Jurassic shallow-
marine sediments are interbedded with volcanic
rocks in the South Shetland Is and Adelaide I,
and gently folded. ?Middle Jurassic plant beds
underlie more silicic lavas near the northeast
tip of the Peninsula (Thompson, 1982). The 8000
m - thick shallow-water volcaniclastic Fossil
Bluff formation, of Oxfordian-Kimmeridgian to

Albian age and exposed on Alexander I. (Bell
1974, 1975), provides the only obvious example
of a well-developed post-Triassic fore-arc
basin (Suárez, 1976) within the Scotia Arc region.
The submerged continental shelf, however, which
narrows only gradually to the north and southwest of Alexander I, may hide other examples,
as is suggested by the subdued nature of aeromagnetic profiles over the area to the north
(Renner et al., 1982).

On the east coast of the Peninsula also, the
Mesozoic sedimentary rocks are much more abundant
in the south. From the southern Black Coast at
73°S around the bend as far as eastern Ellsworth
Land, the major part of the exposed rock belongs
to the Latady formation, a thick, shallow-marine
volcaniclastic sequence of Middle to Late
Jurassic age, interbedded towards the Pacific
with dacitic lavas. The formation is intensely
folded about axes parallel to the Peninsula and
intruded after the time of deformation by calc-
alkaline plutons of ages ranging from 120 to 95
Ma (Williams et al. 1972; Laudon, 1972; Rowley
and Williams, 1982; Farrar et al., 1982). The
similarities in provenance and deformational
style between the Latady formation and the
Yahgan and Cumberland Bay formations of Tierra
del Fuego and South Georgia have been noted
(Suárez, 1976), renewing the suggestion that the
Weddell Sea opened as a back-arc basin behind
the Antarctic Peninsula, analogous to the
Western Pacific back-arc basins of today (Barker
and Griffiths, 1972; Dalziel, 1974). The contrast between Mesozoic sedimentation on the
northern and southern parts of the east coast
of the Peninsula could then result from the
absence and presence (respectively) of a foreland shelf region to prevent wider dispersal
of the sediments and provide a buffer against
which they might be deformed, as with the
Springhill Platform of the Magallanes basin
(Natland et al. 1974). The analogy cannot be
taken too far, however. There is no trace at
the base of the Peninsula of foreland-derived
sediments, an ophiolitic floor to the basin
or its continued development and deformation,
and both the silicic volcanism preceding opening
and the post-deformational intrusion which
defines the latest time of 'closure', appear
to have happened up to 20 Ma earlier there than
in South America.

Weddell Sea magnetic anomalies strike approximately east-west (Jahn, 1978; Barker and Jahn,
1981) and young northward; ages between 40 Ma
in the north and 155 Ma in the south have been
reported (LaBrecque and Barker, 1981), but the
inconsistent shapes of anomalies along strike
make these identifications no more than tentative. In general terms, the early history of
the Falkland Plateau (Barker, Dalziel, et al.,
1976) and magnetic data from the Indian Ocean
(Bergh, 1977; Segoufin, 1978; Simpson et al,
1979) both point to an earlier separation of
East Antarctica from Africa (~150 Ma) than the

start of South Atlantic opening (~130 Ma : Larson and Ladd, 1973), although the older Indian Ocean and Weddell Sea anomalies could not have been produced by the same 2-plate system. However, silicic volcanic activity on the east coast of the Peninsula may have borne the same precursory relationship to Africa-Antarctica separation (and Weddell Sea formation) as the South American activity appears to have done to South Atlantic opening (Dalziel, 1974: Bruhn et al, 1978), at least in time and perhaps genetically.

The Cretaceous and early Tertiary geological history of the Antarctic Peninsula is one of continuing volcanic and plutonic activity accompanied by uplift, as shown by the presence of clasts of plutons and older sediments within the sedimentary rocks on its flanks. Mid- and Late Cretaceous plutons extend over the entire length of the Peninsula and are by far the most numerous in the Scotia Arc region, as elsewhere. Tertiary calc-alkaline igneous activity was confined to the west coast of the Peninsula, as already noted, and appears to have ceased progressively northward, in accord with but significantly ahead of the cessation of subduction at the trench (Barker, 1982 and Section B2). Clearly an appreciation of the way in which magmagenesis responds to changes in the parameters of subduction throws light on the processes involved, and is important to the interpretation of the geological record. There are obvious possibilities of investigating this particular relationship in great detail in the Scotia Arc region, as additional radiometric (particularly Rb-Sr) data become available and offshore magnetic coverage improves. Recent investigations of aspects of this problem (e.g. Saunders et al, 1980; Hawkes, 1981, 1982; Barker, 1982; Pankhurst, 1982b) are described in Section C2.

The most precise offshore data are for the Cenozoic, yet onshore Cenozoic radiometric data are sparse: Palaeocene and Eocene radiometric ages have been obtained for plutons from as far south as 65°S on the west coast of the Peninsula (Rex, 1976) and for plutons and lavas from the South Shetland Is (Dalziel et al, 1973; Grikurov et al., 1970; Watts, 1982). South Shetland lavas of this age are intermediate in composition between basaltic andesites and island arc tholeiites, but this is merely in accord with the compositional gradient across the Peninsula already discussed. The South Shetland Is have an earlier history which is also essentially similar to that of the Peninsula. Post-Eocene calc-alkaline rocks are rare: a 9.5 Ma granodiorite from near Elephant I (Rex and Baker, 1973), an early Oligocene lava from the southeast side of Bransfield Strait (Rex, 1972) and the above-mentioned Anvers I pluton, which Gledhill et al. (1982) use to suggest a Miocene to Recent age for andesitic lavas from northern Anvers I (Hooper, 1962). Young olivine basalts are found on nearby Brabant I (Gonzalez-Ferran

and Katsui, 1970). Among the Eocene volcanics (Watts, 1982) are some thought originally to be Miocene (Barton, 1965).

There was a renewal of volcanism in the latest Miocene, but of very different distribution and range of compositions and associated directly or indirectly with back-arc extension in Bransfield Strait. This development is discussed in detail in Section C1b, being of considerable interest as a possible model for ensialic back-arc basin formation. Here we wish to note merely that it may have been preceded by a period (since the Eocene) when calc-alkaline activity was not extensive, despite the likelihood of continuous subduction.

There is some debate (Baker, 1976; Barker, 1976; Baker et al., 1977; Weaver et al., 1979, for example) about the nature of the connection between Bransfield Strait opening and volcanism on the east coast of the Peninsula. This latter produced alkali olivine basalts and hawaiites, on James Ross I (Nelson, 1975) at first (6.5 Ma ago - Rex, 1976) but later extending north and northeast as far as Paulet I (Baker et al., 1976) and southwest to Seal Nunataks (Fleet, 1968), where radiometric ages of less than 1 Ma have been obtained (Rex, 1976; Baker et al., 1977), and Jason Peninsula (Saunders, 1982). The east coast lavas have no real calc-alkaline affinities, but in view of their approximate association in time and space with Bransfield Strait extension, which is itself almost certainly subduction-related, a connection with events at the Pacific margin seems as likely as an essentially intra-plate origin.

Uplift associated with the east coast volcanism exposed thick (>5 km) sequences of Upper Cretaceous and Tertiary sediments. These were derived from the Peninsula to the west, and probably represent the upper part of a thick clastic wedge occupying the broad but largely unknown Weddell Sea shelf. This speculation is supported by British Antarctic Survey aeromagnetic profiles, which are quiet away from the known exposures of volcanic rock (Renner et al., 1982).

B6. Scotia Sea - Cenozoic evolution

As Figure 1 makes clear, the Scotia Sea exists at present as a complication on the SAM-ANT boundary, so that its development is at least partly dependent upon SAM-ANT motion. This dependence is unlikely to be new, so for a full understanding of Scotia Sea evolution some knowledge of the history of SAM-ANT motion is required. At present this is known directly only over the last 5 Ma and in the extreme east at the Bouvet triple junction (Sclater et al., 1976). SAM-Africa and Africa-ANT motion in combination might extend this knowledge back in time, but the latter is known at present in only one area (Bergh and Norton, 1976) and some of the magnetic anomaly identifications there have been questioned (Norton and Sclater, 1979). It has been suggested (Sclater et al., 1978) that the great length of some of the fracture zones on the S.W. Indian Ridge indicates the longevity of the present spreading regime (for up to 40 Ma at present spreading rates) and the same general argument could be made in respect of the long E-W fracture zones of the SAM-ANT boundary. Rates as well as directions of spreading may alter however, and much better data are required before the history of the simple, notional SAM-ANT motion in the Scotia Sea region can be estimated. Meanwhile, there is the general indication of a recent Indian Ocean reconstruction, when combined with South Atlantic opening (Norton and Sclater, 1979; Norton, 1982), that southern South America and the Antarctic Peninsula remained close to each other from before 110 Ma until after 40 Ma. The uncertainties of such a reconstruction are of course magnified when transposed into the Scotia Sea region. Also, this particular attempt takes no account of possible East Antarctica-Antarctic Peninsula motion since 110 Ma ago. Nevertheless, in classifying the Scotia Sea as a complication of the SAM-ANT boundary, it is perhaps reasonable to assume that it evolved over the past 30 to 40 Ma, when SAM-ANT motion may have been much larger than hitherto.

During the period of the Geodynamics Project there has been considerable progress in establishing the nature of the Scotia Sea and surrounding Scotia Ridge, although their evolution is not yet fully understood. The Scotia Ridge (Figure 6) is composed largely of continental fragments, whose onshore geology is incompatible with their present isolated situation, but indicates an original position close to a subducting continental margin (presumed to be the Pacific margin, between the Peninsula and southern South America). Where the land geology can be extrapolated offshore, structures are typically truncated at the margins of the continental blocks, and the last major event in their evolution appears to have been their fragmentation (Barker and Griffiths, 1972). Thus the batholith and ophiolite complex of the Patagonian Cordillera is truncated abruptly at the margin south and east of Cape Horn, and Burdwood Bank, an eastward continuation of the pre-Cordillera, or Andean foothill belt, appears to require a southerly sediment source where only ocean floor exists today (Barker and Griffiths, 1972; Davey, 1972a). The north Scotia Ridge east of Burdwood Bank comprises three small elevated blocks, then the Shag Rocks block, and finally South Georgia. The origin of the three blocks is unknown, except that their northern margin incorporates a wedge of deformed sediments (Ludwig et al, 1979; Ludwig and Rabinowitz, 1982) similar to but less well developed than that on the northern margin of Burdwood Bank. The Shag Rocks block is largely non-magnetic, and at

least partly composed of rocks resembling those of the South Orkney Is metamorphic complex (Tanner, 1982).

The geological similarities between South Georgia and southern Tierra del Fuego, first noted by early explorers and more recently by Katz and Watters (1966), have since been confirmed in overwhelming detail (see, for example, Dalziel et al., 1975; Suárez and Pettigrew, 1976; Storey et al., 1977; Winn, 1978; Simpson and Griffiths 1982; Storey and Mair, 1982; Tanner, 1982b). The comparison argues in favour of an original position for South Georgia directly east of Cape Horn (and thus south of Burdwood Bank and a likely sediment source). Its subsequent 1600 km eastward translation therefore becomes one of the tests of any hypothesis of Scotia Sea evolution.

In a similar fashion, much of the south Scotia Ridge appears to be composed of continental fragments, which have moved eastward relative to their original positions adjacent to the northeastern end of the Antarctic Peninsula. The South Shetland Is. probably have moved only by virtue of the 30 km opening of Bransfield Strait over the past 1-2 Ma (see Section C1b). The Elephant and Clarence I group may also have moved very little. They expose remnants of a lower arc-trench gap terrain of late Cretaceous or greater age (Tanner et al, 1982), and with an apparent polarity consistent with a location at the Pacific margin of the Peninsula similar to their present one (Dalziel, 1982). The South Orkney Is., at the northern edge of a large continental block, show a similar arc-trench gap assemblage, but of evident pre-M. Jurassic age (Dalziel et al, 1981; Tanner et al., 1982). To the south of these islands a 4 km-deep sedimentary basin gives way to a broad zone of magnetic anomalies (Harrington et al., 1972), which could represent an eastward extension of the Antarctic Peninsula batholith. At present, the South Orkney Is. block is separated from the Peninsula in the south by the Powell Basin, probably oceanic, and in the north by two flat-topped ridges with a narrow, deep intervening trough. The crustal structures of the northern ridge resembles that of Elephant and Clarence Is., buried beneath up to 2 km of unmetamorphosed sediment, while the southerly ridge appears identical to the southern, magnetic part of the South Orkney Is block (Watters, 1972). The original locations occupied by these fragments are likely to be disputed until additional information is obtained about the exact nature of the broad areas of submerged shelf.

The continental origin of the remainder of the South Scotia Ridge is by no means certain: its nature and that of a number of intermediate depth blocks in the southern part of the Central Scotia Sea are best considered in the light of Scotia Sea marine magnetic data. The South Sandwich Is., which form a young intra-oceanic volcanic

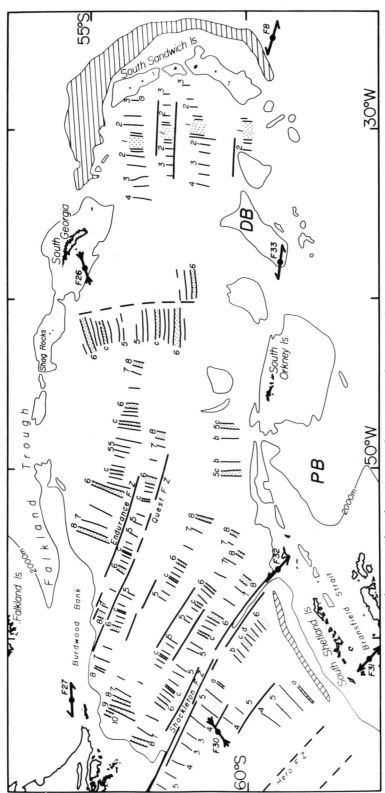

Fig. 6. Ocean floor magnetic anomalies in the Scotia Sea province compiled from Barker (1972b), Barker and Burrell (1977) and LaBrecque and Rabinowitz (1977). Anomalies northwest of the South Orkney Islands are only tentatively dated (Hill and Barker, 1980). Preferred interpretations of earthquake focal mechanisms from Forsyth (1975). PB is the Powell basin and DB Discovery bank.

island arc associated with subduction of oceanic crust belonging to the South American plate at the east-facing South Sandwich Trench, are also best understood in the context of Scotia Sea evolution.

Oceanic magnetic lineations belonging to four distinct spreading systems have now been found within the Scotia Sea east of the Shackleton fracture zone (Figure 6). Most of the western area, including nearly all of Drake Passage, was formed between 28 Ma and 6 Ma by the separation of the South Scotia Ridge from southernmost South America, approximately along 120°T (Barker and Burrell, 1977). Spreading was fast initially (~46 mm/year until 26 Ma) but slowed progressively; it may not have stopped at exactly 6 Ma, but a model using 20 mm/year fits the magnetic anomalies well between 16 Ma and 8 Ma, and produces the observed amount of younger ocean floor (in the region close to the Shackleton fracture zone, at least) if extrapolated to 6 Ma. The spreading centre coincides with an unsedimented, dissected double ridge lying nearly centrally in Drake Passage (Barker, 1970, 1972a; Ewing et al, 1971).

Anomaly 8 lies at the continental margin in some places but not in others, so that an earlier extensional episode is implied. A9 and A10 are seen outside A8 on some profiles in the southeast (Barker and Burrell, 1977), and northeast of Cape Horn (LaBrecque and Rabinowitz, 1977), but not elsewhere, and it seems that a considerable reorganisation of spreading, involving ridge jumps, may have taken place just before A8 time (~28 Ma). Figure 7, a speculative reconstruction to A8 time, shows the distribution of older ocean floor. The positions of the Shag Rocks (SR), South Georgia (SG) and South Orkney (SO) blocks are only schematic, but do illustrate how the

eastward dispersal of blocks originally in a compact grouping at the Pacific margin may have proceeded. The reconstruction in Figure 7 cannot be made if only two rigid plates were involved in Drake Passage opening, and Barker and Burrell assumed that for part of the time the southeast side of Drake Passage formed two separate plates. More recent data appear to bear this out (see below).

The rapid east-west back-arc extension which formed the eastern Scotia Sea over the past 8 Ma (Barker, 1972b) is described in some detail in Section C1a, because of its geochemical interest and possible significance for the study of mechanisms of back-arc extension. Sandwich plate motion is much faster than, and virtually independent of, major plate motion; the extension complements subduction at the trench (with the addition of the north Scotia Ridge component of SAM-ANT motion, which is considerably slower). This independence may be contrasted with Drake Passage opening which, from 16 Ma until about 6 Ma, was similar in direction and rate to present-day SAM-ANT motion. Although the history of SAM-ANT motion is unknown, it is thus possible that the Drake Passage spreading system was originally part of a much simpler SAM-ANT boundary. Clearly, the existence of South Sandwich-like subduction removes the need for a close coupling between Drake Passage opening and SAM-ANT motion (as at present - see Figure 1) and instead converts it into something resembling back-arc extension, which in turn raises the problem of the origin of the intervening, central part of the Scotia Sea.

The Central Scotia Sea is more elevated than Drake Passage, with thicker sediment cover, and its crustal structure (Allen, 1966; Ewing et al. 1971) resembles that of back-arc basins. DeWit (1977) argued on these grounds that it was of Early Cretaceous age, a broad eastward extension of the Rocas Verdes ophiolite complex in southern South America and South Georgia (see Section C1c). Barker (1970) and Barker and Griffiths (1972) suggested that the Central Scotia Sea could have opened at the same time as Drake Passage, essentially as a species of back-arc extension behind ancestors of the South Sandwich arc and trench. Magnetic anomalies in the Central Scotia Sea however have proved difficult to map, and it is only recently that either suggestion has gained any support (Hill and Barker, 1980).

A reconnaissance of the elevated blocks of the eastern South Scotia Ridge (Barker et al., 1982) has shown that two of them, including the largest (Discovery Bank, DB in Figure 6) are composed of primitive island-arc tholeiites, virtually identical to South Sandwich Is rocks but of early to mid-Miocene age (12-20 Ma). This 'Discovery arc' predates the South Sandwich arc (which probably rests on ocean floor only 5-7 Ma old: Barker, 1972b) and, where sampled, is no longer active. Its original lateral extent is uncertain. The sampled blocks lie within an

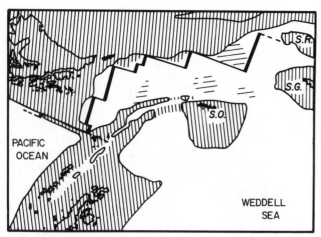

Fig. 7. Reconstruction of Drake Passage at Anomaly 8 time (~28 Ma) from Barker and Burrell (1977). Positions of Shag Rocks (SR), South Georgia (SG) and South Orkney (SO) continental fragments are speculative. Note asymmetric position of ridge crest within Drake Passage.

area which at present is being dissected by Scotia-ANT motion (Figure 1) and it is not known how many, if any, of the adjacent, unsampled blocks have a similar origin. If South Sandwich back-arc extension over the last 8 Ma is taken into account, the line of the present South Sandwich arc would have formed a northward extension of Discovery Bank, and although the present arc is too young, remnants of an earlier Discovery arc may possibly occupy parts of the South Sandwich arc-trench gap. Such continuity of arc and of subduction might provide an explanation of the change from Discovery to South Sandwich arcs. If a more southerly ridge crest section of the southwest Atlantic spreading system than now exists migrated into the southern part of the Discovery trench (as will occur again at the South Sandwich trench in about 5 Ma from now - see Figure 1), subduction along that part might have stopped, while the weight of the subducted slab would have caused it to continue sinking along the more northerly (S. Sandwich) part (Barker et al., 1982; Barker and Hill, 1981).

The detection of early mid-Miocene subduction in the eastern South Scotia Ridge has strongly influenced the interpretation of east-west magnetic anomalies recently mapped in the Central Scotia Sea (Hill 1978; Hill and Barker 1980). At present, a correlation is preferred with that part of the magnetic reversal time scale (LaBrecque et al., 1977) lying between 21 and 6 Ma, which is clearly compatible with an origin for the Central Scotia Sea by extension behind the Discovery Arc. However, that section of the time scale is notoriously ambiguous (see for example Tomoda et al., 1975; Watts and Weissel, 1975 and Murauchi et al., 1976, on spreading in the Shikoku Basin). Also, the Central Scotia Sea sequence is short and anomaly shapes are not always consistent between tracks, so that some uncertainty about its identification remains. Strong similarities with the Drake Passage anomaly sequence, agreement with the oceanic age-depth curve and the general sediment distribution, however, all support a Neogene age for the Central Scotia Sea rather than the Early Cretaceous age suggested by DeWit (1977).

An even greater uncertainty attaches to the identification of the north-south magnetic lineations in the small 'Protector' basin northwest of the South Orkney Is. (Hill and Barker, 1980) because each side of the symmetric anomaly sequence is only 120 km long. Tentatively, they have been dated as 13-16 Ma old. Whatever the age of this sequence, however, the lineations introduce the attractive concept of repeated small back-arc extensional basins separating all of the elevated blocks in the southern part of the Central Scotia Sea, and make it easier to see how, as Barker and Burrell (1977) suggested, the southeastern side of Drake Passage might have formed two separate plates for part of its history.

It is difficult to assemble a definitive model for the evolution of the Scotia Sea until at least some of these uncertainties have been resolved. Nevertheless, there are clear indications that most of the Scotia Sea formed over the past 30 to 40 Ma, which was suggested on more general grounds at the start of this section. Furthermore, it can be seen that, perhaps because of the small size of the region, no simple tectonic process persists unchanged within it for very long; change itself is one of the region's most prominent characteristics. This may provide a satisfactory viewpoint from which to examine present plate motions and boundaries around the Scotia Sea, in which only a small fraction of the tectonic systems described above are actively involved.

Sparse earthquake activity (Figure 1) occurs along both north and south Scotia Ridges, the Shackleton fracture zone and Bransfield Strait and on spreading centres in western Drake Passage and the eastern Scotia Sea. With other data, the earthquakes define four microplates, the Drake and Shetland plates (considered in detail in Sections B2 and C1B), the Sandwich plate (Section C1a) and the Scotia plate which, although bigger than the other three combined, is the least known. Its long boundaries, the north and south Scotia Ridge, are sub-parallel to the east-west direction of the slow, sinistral SAM-ANT motion which is partitioned between them. First motion studies by Forsyth (1975) on a small number of earthquakes from these boundaries (Figures 2 and 6) indicate that motion along both is also sinistral (not an inevitable situation: cf. the Caribbean), and therefore also slow and probably approximately east-west.

The infrequency of earthquakes and uncertainties in their location make it difficult to see exactly where the boundaries lie. Barker and Griffiths (1972) suggested that one followed the deep narrow trough which bisects the south Scotia Ridge between 40° and 55°W, and which widens where it steps southward at 51°W, which is consistent with sinistral motion. The shape of the trough is not well-mapped in detail, and the boundary farther east is totally unknown, but Barker et al. (1982) argue that Scotia-ANT motion was and is dissecting the dead Discovery arc, so that an additional short, ridge-crest offset may occur in the vicinity of 35°W. The boundary seems capable of yielding a pole of Scotia-ANT motion if mapped in detail but, if motion is slow, magnetic survey may not reveal a spreading rate for the narrow ridge-crest section, as in the similar Cayman Trough (Macdonald and Holcombe, 1978).

The Shackleton fracture zone is also presumed to be the locus of pure strike-slip, but may be more complex. At its southern end it forms part of the Scotia-Shetland and Scotia-Drake boundaries. Farther north it merges with the Pacific margin of southern Chile as part of the

Scotia-ANT boundary, which probably becomes obliquely subducting as it curves northward, towards a triple junction at the western end of the Straits of Magellan. This would appear to be a TTF junction, stably configured, but other possibilities have been proposed, influenced perhaps by the complexity of the onshore geology and of fault-controlled mid-slope basins offshore (Forsyth, 1975; Fuenzalida, 1976; Herron et al., 1977; Winslow 1982 and Figure 5). Certainly there is much evidence onshore (see Section B4) that the present configuration has not been in existence for very long. The fault boundary at the TTF junction is oriented at about 120°, lies along the Straits of Magellan and is sinistral (Winslow, 1982). It forms the western end of the SAM-Scotia boundary, which eastward takes a line presumed by most workers, in the absence of epicentral evidence, to lie along the Falkland Trough on the northern flank of the North Scotia Ridge. The presence of a vertical component on Magellan Strait earthquake first motions (Forsyth, 1975) and the observation that older structures there are being reactivated (Winslow, 1980) makes the use of this lineament in the computation of SAM-Scotia plate motions open to question. Farther east, first motions are much more nearly east-west, and therefore essentially strike-slip. Whether the accretionary wedge found on the northern flank of the North Scotia Ridge (Ludwig and Rabinowitz, 1982) could have grown as a result of such strike-slip motion within a narrow elongated sedimentary basin, or requires an earlier phase of motion with a much greater component of north-south convergence, is uncertain.

Thus, neither SAM-Scotia nor Scotia-ANT motions are known exactly at present, and their precise computation presents difficulties, essentially because both are probably slow. There is indirect evidence, however, that neither of the present, Scotia Ridge boundaries has been active in its present sense for very long. They may have started only when Drake Passage opening stopped, at about 6 Ma.

C. Geodynamic Processes

C1. Back-arc extension

Cla Sandwich Plate. The South Sandwich island arc sits on the small, D-shaped Sandwich plate, whose rapid eastward movement is almost completely independent of major plate motions. Subduction of South Atlantic ocean floor from the SAM plate at the South Sandwich trench is compensated by fast back-arc extension at about 30°W in the East Scotia Sea, plus whatever portion of the much slower SAM-ANT motion occurs along the North Scotia Ridge (Figures 1 and 6). The East Scotia Sea provided the first and still possibly the best example of normal (i.e. Vine-Matthews) spreading behind an island arc (Barker, 1970, 1972b;

Barker and Hill, 1981). Anomalies out to about 8 Ma can be seen on the western flank of the spreading centre, but in the east the volcanic islands themselves lie on ocean floor which is only 5-7 Ma old, and was generated during the present spreading episode. Spreading is accelerating, from 50 mm/year 2 Ma ago to 70 mm/year at present, with a slight asymmetry favouring accretion to the eastern, trench side, but there is no sign of such systematic north-south variations as might result from continued bending of the arc, or a near pole of opening. The exact start of spreading is difficult to detect. North-south magnetic lineations are not seen beyond 8 Ma and in places the topography becomes more elevated westward, with repeated east-facing scarps, but not everywhere and not on an isochron. Barker et al. (1982) argue that subduction was taking place before 8 Ma, but may have been southeast-directed and was causing north-south back-arc extension (Section B6). The transition to east-west extension may have been somewhat untidy (Barker and Hill, 1981).

Relocated earthquakes (Brett, 1977) show that the dip of the Benioff zone steepens from 45°-55° in the south and central areas to near-vertical at the northern end. Forsyth (1975) used first motions of earthquakes to suggest that the descending slab is being torn at its northern end, at depths of 50 to 130 km, to accommodate the rapid eastward migration of the Sandwich plate. Frankel and McCann (1979), however, attribute the same seismicity to stresses <u>within</u> the subducting plate as it adjusts to the sharp curvature of the NE corner of the trench, since some of the earthquakes do not show a fault plane parallel to the Sandwich-SAM slip vector.

Only one earthquake from beneath 180 km has been observed (at 250 km - Brett 1977) and the first motions of intermediate depth earthquakes show down-dip extension at the northern end (away from the tearing zone) but down-dip compression in the centre and south (Isacks and Molnar, 1971; Forsyth, 1975). Forsyth (1975) and Brett (1977) suggest that this contrast (together with the steeper dip) arises from the greater age of the oceanic lithosphere being subducted at the northern part of the trench; the older, denser material in the north would be pulling down with it the younger, less dense material in the south. The age of the subducted ocean floor is unknown, but if it had formed at the spreading centre as configured today, the long ridge offset would produce the suggested age gradient (see Figure 1). A difference in buoyancy of the subducted slab, stemming from the same contrast in age, has been invoked by Frankel and McCann (1979) to explain the more elevated southern section of South Sandwich fore-arc. However, the model of the early evolution of the Sandwich plate produced by Hill and Barker (1981) implies a longer and more complicated history for the southern than the northern part of the South Sandwich arc and trench, which provides an alternative explanation.

Considering how much oceanic lithosphere has been produced in the back-arc basin during the past 8 Ma, the deepest earthquakes in the Benioff zone are not very deep (Barker, 1972b), even including the 250 km event. As Brett (1977) commented, this could be explained by earlier resorption of younger oceanic lithosphere into the mantle (Vlaar and Wortel, 1976) provided that earthquakes in the north occurred at substantially greater depth than in the south, which at present is not obviously the case.

Earthquakes from the Benioff zone have also been used to examine the back-arc region. Barazangi et al., (1975) failed to find the expected zone of high attenuation using pP waves at teleseismic distances, but were not able to sample areas younger than 3-4 Ma. Brett and Griffiths (1975) however, using arrivals at a local station (KEP at 5 to 8 degrees), found pronounced low-velocity, low-Q characteristics within the back-arc region.

There has been much discussion over the past few years about the causes of back-arc extension (see, for example, Molnar and Atwater, 1978). Chase (1978b) has used the Sandwich plate example to consider the significance of absolute plate motions. Using the hotspot reference frame and reasonable values for relative plate motions (Chase, 1978a) he shows that almost all trenches are 'advancing' in absolute terms, so that the subducting slab sinks vertically and is torn back, rather than merely sliding along its length. Moreover, at the four regions (including the Sandwich plate) where rapid back-arc extension has now been convincingly demonstrated, the plate containing the remnant arc (here the Scotia plate) is 'retreating', so that without back-arc extension the trench would not advance. Chase refers this mechanism to the 'trench suction' of Forsyth and Uyeda (1975). Despite the attractiveness of this synthesis there are problems; if all trenches are advancing then the mantle beneath the entire descending slab must be being displaced, but to where? This is the very region, beneath the asthenosphere, whose worldwide fixity is the basic assumption behind the validity of the hotspot reference frame, on which the study is based. Barker and Hill (1980) use the East Scotia Sea and other back-arc basins to argue a 'thermal' model for asymmetric spreading. Their conclusions also provide some indirect support for Chase's (1978b) model of back-arc extension, as against those, such as the 'forced convection' model of Toksoz and Bird (1977) which are not related to absolute plate motion.

With the exception of Leskov I., the South Sandwich Is. lie on a single arcuate axis, 100 to 140 km above the top of the descending slab (Brett, 1977). Volcanic or fumarolic activity has been detected on most of them, and radiometric ages of 0.7 and 4.0 Ma have been obtained from rocks thought to be relatively old (Baker, 1968, 1978), which accords with their situation on oceanic crust probably only 5 to 7 Ma old (Barker, 1972b). Most of the rocks sampled are primitive island arc tholeiites (Jakes and Gill, 1970), but calc-alkaline andesites do occur, most notably on Leskov I. which lies about 60 km west of the main arc and the deepest earthquakes, on ocean floor about 5 Ma old. Baker (1978) suggests that the higher potash content of the Leskov I. lavas indicates its situation over a deeper part of the descending slab (undetected otherwise) and that this occurrence marks the beginning of a later stage of island arc volcanism in the South Sandwich Is.

As has already been mentioned, an intra-oceanic arc ancestral to the South Sandwich arc has been proposed on the basis of the very close similarity of rocks dredged from part of the eastern South Scotia Ridge to the arc tholeiites mentioned above (Barker et al., 1982). Named the Discovery arc, it was probably directed southeastward, and active over at least the period 20 to 12 Ma. Barker et al. (1982) and Barker and Hill (1981) explain its evolution in terms of ridge crest/trench collisions.

Saunders and Tarney (1979) have analysed rocks dredged from four widely separated sites on the back-arc spreading centre (Saunders et al. 1982b) and used the results to consider the possibilities of distinguishing between mid-oceanic ridge and back-arc basin origins for ophiolite complexes. The rocks were compositionally similar at any one site, but between sites ranged from sub-alkaline olivine tholeiite to highly vesicular quartz-normative basalt. Modelling suggests that, with respect to the incompatible elements Ti, P, Zr, Hf, Nb, Ta, Y and the light REE (Hawkesworth et al., 1977), rocks at all four sites could have been derived from essentially the same mantle source by different degrees of partial melting (generally 9-15% but up to 30% for the more primitive of the highly vesicular basalts). In other respects, however, the mantle source is variable, notably in correlated enrichment in K, Rb, Ba and $^{87}Sr/^{86}Sr$ ratio. Saunders and Tarney see similar patterns of variability and enrichment in basalts from other intra-oceanic back-arc basins, and note similarities between the more enriched samples and island arc tholeiites. The variations lie within the range of variability of oceanic ridge samples, but with a tendency towards generation from a less-depleted mantle source. Since this could be produced by the addition of K, Rb, Ba and more radiogenic Sr by dewatering of the subducted slab (apparently locally rather than uniformly), conditions of magma generation could approach those of arc tholeiites. Thus, although an unequivocal distinction between an oceanic or back-arc origin may not be possible for all

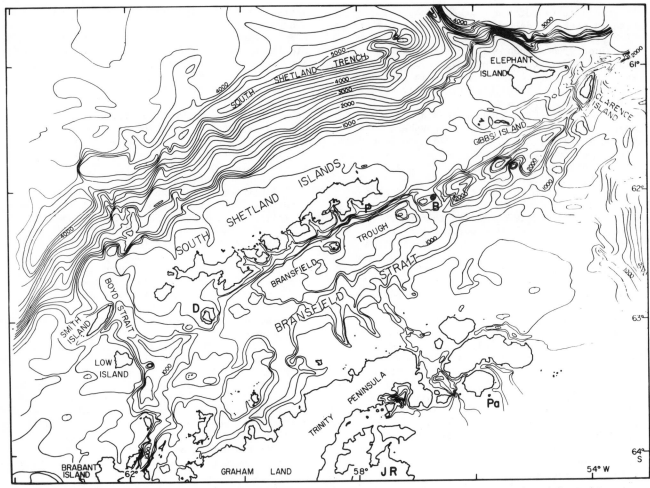

Fig. 8. Bathymetry of Bransfield strait (Davey 1972b), with locations of Deception (D), Bridgeman (B), Penguin (P), James Ross (JR) and Paulet (Pa) Islands.

ophiolite complexes, those such as Troodos which show some island arc characteristics may well have formed in a back-arc environment.

C1b. South Shetland Is. and Bransfield Strait. Interest in Quaternary volcanism in Bransfield Strait and recognition of its unusual nature is of long standing (Hawkes, 1961; Gonzalez-Ferran and Katsui, 1970, for example), and at the start of the Geodynamics Project was intensified by the renewed activity of 1967-70 on Deception I. (Baker et al., 1975). The more important developments over this period, however, are probably the perception of the back-arc extensional environment of Bransfield Strait volcanism and thus of the area's possible value as a model for ensialic back-arc extension elsewhere.

As outlined in Section 2, Bransfield Strait separates the Antarctic Peninsula from the South Shetland Is. along the only part of the Pacific margin of Antarctica where subduction did not end with the migration of a ridge crest

into the trench (Barker, 1970, 1976; 1982, Herron and Tucholke, 1976). Subduction was totally coupled to spreading in western Drake Passage to the north, which produced between 40 and 64mm/year of new ocean floor from 20 Ma to about 4 Ma, and then virtually stopped. There is no earthquake activity associated with the subducted slab at present, and it is generally assumed (e.g. Barker, 1976) that subduction also virtually stopped at 4 Ma.

Beneath Bransfield Strait lies a 2km deep trough, with a steep northwestern slope up to the South Shetland Is. and a more gradual southeastern margin towards the Antarctic Peninsula shelf (Figure 8). The flat bottom of the trough is interrupted by a series of submarine peaks, on a line running through Deception I and Bridgeman I, which is also volcanic and probably very young (Gonzalez-Ferran and Katsui, 1970). Penguin I., an uneroded scoria cone and the only other example of Recent subaerial volcanism, lies off this

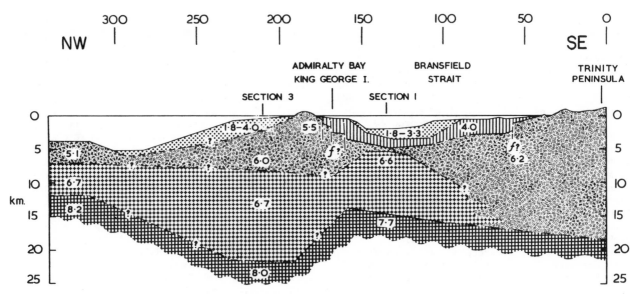

Fig. 9. Crustal structure across Bransfield Strait, from seismic refraction measurements (From Ashcroft 1972).

axis south of King George I, on top of the steep, northwest slope of the trough.

The South Shetland Is and Antarctic Peninsula have a nearly identical geological history extending at least as far back as the Jurassic (see Section B5) and the platforms on which they stand both have a continental crustal structure (Ashcroft, 1972). In contrast, the structure beneath the trough of Bransfield Strait resembles that of an oceanic ridge crest, but with a thicker main crustal layer (6.5 to 6.9 km/sec) overlying 7.6 to 7.7 km/sec mantle at about 14 km below sea level (Ashcroft, 1972 and Figure 9). Normal faulting on the southeastern margin of the South Shetland Is., of uncertain age but associated with the youngest, ?Pliocene to Recent volcanics exposed there (Barton, 1965; Weaver et al., 1979) continues offshore to form the steep northwestern margin of the trough (Ashcroft, 1972). Similar normal faulting in a broader zone underlies the sediments on the more gradual southeastern margin of the trough (Davey, 1972b; Ashcroft, 1972). On these grounds it has been suggested that Bransfield Strait is extensional (Ashcroft, 1972; Barker and Griffiths, 1972; Davey, 1972b), with an initial phase of repeated normal faulting giving way to the generation of a new, near-oceanic crustal section, and is related to subduction at the South Shetland trench. This interpretation raises a different set of questions, of considerable geodynamic interest. What has been the exact timing of the extensional process, given that subduction at the trench probably stopped 4 Ma ago? Does the back-arc extensional model permit any more fundamental understanding of the unusual chemistry of recent volcanism in Bransfield Strait? In view of controversy over the nature of back-arc extension, how close a

resemblance is there between the process of extension in Bransfield Strait and that of the main ocean ridges?

A combination of radiometric dating and detailed mapping may provide an age for the normal faulting on the southeast flank of the South Shetland Is., although samples collected so far (Weaver et al., 1979 and personal communication) have proved unsuitable for dating. Alternatively, the large positive magnetic anomaly associated with the line of submerged peaks occupying the centre of the trough offers the possibility of both dating the extension and examining its nature. The anomaly can be fitted approximately by assuming uniform normal magnetisation of this axial topography (Barker, 1976). If the volcanic activity on Deception I and the shallow earth-quake activity within the Strait indicate that extension is still taking place, then a simplistic interpretation is that the axial trough has formed by straightforward sea-floor spreading over, at most, the past 0.69 Ma of the Brunhes normal magnetic polarity epoch. Clearly, however, any process of extension, if confined within these time limits and to the trough axis, would produce a similar result. Barker pointed out also that a discrepancy between the 1 km of sediments seen by seismic refraction shooting (Ashcroft, 1972) and the 100 m penetration of the reflection profile could mean that lavas or sills were becoming interbedded with the sediments at some distance from the eruptive centre, and possibly offsetting the effects of reversely magnetised lavas on the flanks. More recently Roach (1978 and personal communication) has interpreted a detailed magnetic survey of the Bransfield trough. He concludes that, in many parts of the Strait, the introduction of narrow strips of reversely magnetised material along the flanks of the axial trough is the most reasonable way of obtaining a precise fit to the measured anomaly. The degree of disorder which is observed, particularly, on the flanks, he attributes to the immaturity of the spreading system, and considers that Bransfield Strait has opened by classical (Vine-Matthews) ocean floor spreading rather than by some less-ordered alternative, such as the diffuse extrusion suggested by Karig (1971). Under this hypothesis, about 30 km of extension has taken place, at 24 mm/year for 1.3 Ma. Fracture zone orientations are difficult to detect over so short a distance, but presumably will not differ greatly from the 315°T azimuth of least compressive stress obtained for a Bransfield Strait earthquake by Forsyth (1975: see Figure 2).

This chronology, if accepted, is of considerable geodynamic interest. There is a gap between the time of cessation of Drake Passage spreading (and therefore of fast subduction) at about 4 Ma, and the onset of spreading at 1.3 Ma. The faulting seen onshore and associated with the steep northwest margin of the trough is not precisely dated but, in that its associated calc-alkaline volcanism is at least partly Pliocene in age (Barton, 1965; Weaver et al., 1979), it becomes more reasonably precursory to the extension, although not necessarily as old as 4 Ma. All of this is consistent with the thesis that the cause of extension was the cessation of subduction. If, as suggested in Section B2, spreading in western Drake Passage stopped because the westward propagation of east-west compressive stress (Forsyth, 1975) caused the long, strike-slip boundaries of the subducting plate to seize, then even though the subducted part of the plate was not very old (≳16 Ma, Barker, 1970, 1972a), its excess mass would cause it to continue to sink. This 'trench suction' (Forsyth and Uyeda, 1975) would result in the northwestward migration of the trench and thus in extension in Bransfield Strait. Conversely, of course, extension in Bransfield Strait over the past 1.3 Ma implies renewed subduction (although no associated earthquake activity is observed) and the existence of a separate 'Shetland' microplate (Figure 1).

The volcanic rocks of Deception, Bridgeman and Penguin Is. mark a major departure from a pattern of igenous activity along the Antarctic Peninsula which had persisted since the mid-Jurassic. Deception I in particular is much studied and the unusual chemical nature of its volcanism has long been recognised. In this report we concentrate on more recent work (for example Weaver et al., 1979) which has had the advantages of both the improved understanding of tectonic context, as outlined above, and a wider range of analyses, including trace element and isotopic data.

Deception I is a composite stratovolcano, with lavas ranging from basalt to rhyodacite. There appears to have been a trend to more evolved, acid compositions with time, but basaltic magma was available throughout. Bridgeman I lavas are mainly basaltic andesite, with subsidiary basalt, and mildly alkaline olivine basalts are found on Penguin I (Gonzalez-Ferran and Katsui, 1970; Baker et al., 1975; Weaver et al., 1979). The geochemistry is confusing, displaying some characteristics of mid-ocean ridge basalts (MORB) and others more usually associated with island arc magmatism. In particular, the lavas resemble MORB in having high Na/K ratio and low K and Rb (Penguin I. less so), but have higher $^{87}Sr/^{86}Sr$ (0.7035 to 0.7039) and Ce_N/Yb_N ratios than MORB; alkali-silica and AFM plots suggest some calc-alkaline affinity. Many of the major-trace element and trace-trace element relationships typical of MORB are decoupled in Bransfield Strait rocks, and the Rb/Sr - $^{87}Sr/^{86}Sr$ systematics appear to have been upset, probably reflecting the complexity of a mantle source which has underlain an island arc for at least 180 Ma. The range of Deception I lava compositions can be accounted for by fractional crystallisation, and Weaver et al., (1979) strengthen the link with MORB by characterising the rhyodacite as the subaerial equivalent of oceanic plagiogranite. Bridgeman and Deception Is. lavas appear to have had a similar mantle source, except for possible enrichment of the former locally by dewatering of the downgoing slab. The source for Penguin I lavas appears however to have been different, probably deeper and garnet-bearing. The lack of cohesion in even the simplified analysis described above is disappointing. Of course it may merely be reflecting inevitable mantle heterogeneity, but there are other possibilities. It has been suggested, for example, that recently-erupted Deception lavas originated beneath its northern end (Baker et al. 1975), which would be significantly 'off-axis'. It would be useful therefore to sample some of the submarine volcanoes lying along the axis, to see if some of the apparent variability can be eliminated.

Clc. Southernmost Andes - Rocas Verdes Complex. Detailed field study in the southern Andes over the past few years has confirmed the suggestion first put forward by Katz (1973), that the rocks previously mapped by Chilean geologists as the 'Rocas Verdes' (green rocks) can be interpreted as the quasioceanic portion of the floor of a marginal basin. This basin opened in the latest Jurassic or earliest Cretaceous behind or within a magmatic arc active along the Pacific margin of South America (Dalziel et al., 1974). The detailed study of the Rocas Verdes revealed that in several places most of the pseudo-stratigraphy of an ophiolite complex is present (Figure 10). Admittedly, no ultramafic rocks have been found, but most probably this is the effect of exposure level. Only a few hundred metres of gabbro can be seen above

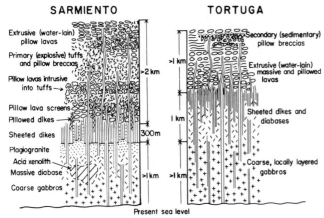

Fig. 10. Ophiolite pseudo-stratigraphy as displayed within the Sarmiento and Tortuga complexes (for location see Fig. 3). From DeWit and Stern (1978).

sea level. The original width and depth of the marginal basin and the proportion of mafic versus continental crust are still the subject of debate (for a critical review of the evidence, see Dalziel, 1981). The basic concept, however, has not only survived, but proved most useful in understanding the Mesozoic-Cenozoic evolution of South Georgia as well as southernmost South America (Dalziel et al., 1975; Dalziel and Bruhn, 1977; Bell et al., 1977; Suárez and Pettigrew, 1976; Tanner 1982b).

The two main ophiolitic complexes identified to date as segments of the floor of the marginal basin in South America, the Sarmiento complex at 52°S latitude and the Tortuga complex at 55°S (Figure 3), have proved to be of global significance for several reasons. Firstly, they are almost unique in being clearly identifiable on the basis of their field relations as marginal basin floor ophiolitic complexes rather than obducted slices of the crust of a major ocean basin. Secondly, they are beautifully exposed, and thirdly, mafic material of most interesting chemical composition has been collected from the Tortuga complex. In reviewing recent advances in our knowledge of the marginal basin we will consider here the magmatic products, firstly the pre-basin volcanism of the Tobifera formation, and then the ophiolite complexes. Section C3, below, deals with deformation and the destruction of the basin.

Following uplift and erosion of the pre-Middle or -Late Jurassic basement complex, there was widespread extrusion of subaerial and shallow marine volcanics of dominantly rhyolitic composition throughout southern South America in the Late Jurassic. Limited radiometric data, supported by field relations, indicate the presence of sub-volcanic granitic plutons within the Andean Cordillera (Halpern, 1973; Nelson

et al., 1980). The association of widespread regional extensional faulting with the volcanicity has led Bruhn and others (1978) to refer to the whole of southern South America in the Late Jurassic as a volcano-tectonic rift zone.

The dominantly silicic composition, large volume and widespread occurrence of the Upper Jurassic volcanics suggest that they resulted from crustal anatexis, rather than magmatic differentiation from some now unexposed mafic rock association. Moreover, geochemical data indicate that the silicic volcanics do not lie along the chemical trends with increasing SiO_2 content which occur in either the plutons of the calc-alkaline Patagonian batholith or the tholeiitic Rocas Verdes (Bruhn and others, 1978). Rather, the Upper Jurassic volcanics are geochemically similar to the rhyolitic rocks of the Taupo volcanic zone in New Zealand, which are thought to be derived by anatectic melting of continental crust (Ewart and others, 1968). The occurrence of calc-alkaline rocks within the volcanics along the Pacific margin in Tierra del Fuego (Bruhn and others, 1978) does suggest that subduction beneath the South American plate was active in the latest Jurassic or earliest Cretaceous prior to the opening of the marginal basin, but the proximity of the South Atlantic basin, which began to rift at about the same time, makes it difficult to conclude that the heat source which produced the Tobifera Formation over such a broad area of South America was solely subduction-related. Most of the calc-alkaline material associated with the Tobifera, closer to the Pacific margin, became separated from that margin when the ophiolite complexes were emplaced, and may thus be regarded as part of a 'remnant arc' in the sense of Karig (1972), despite the absence of evidence concerning the original topography (Section C3).

Mafic rocks along the fringes of the Rocas Verdes terrain in southern Chile intrude the basement complex, the Upper Jurassic volcanics, and early granitic plutons. The main mafic bodies consist of the upper part of ophiolite suites, gabbros, sheeted dykes and pillow lavas, with minor amounts of plagiogranite and associated silicic dykes. Many of the rocks are altered. The metamorphic grade increases from zeolite or greenschist facies in the pillow lavas to amphibolite facies in the gabbros. Metamorphism appears to be related to the hydrothermal convective systems operating at spreading centres during basin formation (DeWit and Stern, 1976, Elthon and Stern, 1978). Geochemically the rocks have affinities with mid-ocean ridge basalts, but K, Rb and Ba contents and Ba/Sr and Ce/Yb ratios are higher and K/Rb ratios are lower in the least altered rocks than in mid-ocean ridge basalts (Saunders et al., 1979).

Much detailed work remains to be undertaken on both of the major ophiolite complexes discovered to date in the marginal basin terrain. Already, however, studies of these rocks have made contributions in several areas. Firstly, it was recognised by Tarney and others (1976) that the overall structure and tectonic development of the Rocas Verdes marginal basin is akin to that of Archaean greenstone belts. Despite the absence of komatiites and slight geochemical differences, this observation led to a model for the evolution of Archaean greenstone belts that has played a part in the recent re-evaluation of Archaean tectonics (Windley, 1976). Secondly, Elthon (1979) has analysed dykes from the Tortuga complex with MgO content as high as 17.61%. These high magnesia dykes are comparable to olivine basalts from Baffin Island and Svartenluk. Elthon argues that whereas these primary liquids are capable of generating the petrologic features of the oceanic crust, this is not possible for tholeiitic basalts with 9-11% MgO which are highly fractionated and that the liquid line of descent is from primary mantle melts with 18% MgO through Ca-rich picrites towards oceanic tholeiites. This in turn argues against the primary nature of oceanic tholeiites. In fact, the spectrum of bulk chemical compositions of basalts in the Tortuga complex is equal to that obtained from work in the FAMOUS area or the entire spread of ocean floor basalts (D. Elthon, personal communication, 1979). Finally, studies of rocks from the Sarmiento and Tortuga complexes by workers at Lamont-Doherty has yielded models for ocean floor igneous processes, metamorphism, seismic layering and even magnetism (DeWit and Stern, 1976, 1978; Stern and others, 1976; Elthon and Stern, 1978).

C2. Arc magmatism - Antarctic Peninsula

The chief influences on magma genesis at subducting margins are generally thought to be the nature (age, sediment cover) of the descending slab, its motion, both with respect to the overiding plate (i.e. convergence rate, dip) and in absolute terms, and the thermal and chemical nature of the crust and mantle above the Benioff zone (i.e. the previous history of subduction). The nature and relative importance of these influences are not understood, and it seems likely that studies of many areas and many subduction episodes will be required before most of the answers are obtained. Nevertheless, a sufficient concentration of research effort into one area may well yield important results, and it seems that parts of the Scotia Arc region have this potential. As has been noted, the incompleteness of the ocean floor spreading record restricts detailed examination of single subduction episodes to those which are relatively recent; the Scotia Arc region contains one such example, a northward-migrating cessation of spreading (Barker, 1982), but our present state

of knowledge, both onshore and offshore, is not yet adequate for its full exploitation. However, more generalised studies of the distribution of subduction-related igneous activity in time and space are possible, and the progress of one such is reviewed here.

The systematic change along the SE Pacific margin in the geographic distribution of igneous activity with time has been noted in Section B2. The steady eastward migration of activity (away from the trench) in northern Chile, between the Early Jurassic and the Oligocene (Farrar et al., 1970), gives way in southern Chile to a zone in which no systematic trend can be seen over the past 180 Ma (Halpern, 1973; Halpern and Fuenzalida, 1978). In the southernmost Antarctic Peninsula only one broadly distributed intrusive phase, covering the period 120-95Ma, has been detected (Farrar et al., 1982), and the orogenic deformation of older dacitic lavas makes their geochemistry of doubtful value.

The main part of the Antarctic Peninsula and its offshore islands possess volcanic and plutonic rocks with a wide range of compositions and ages, and a partly-correlated space-time distribution. Rocks within the age range 185-150 Ma are abundant on the east coast but rare elsewhere, the period 145 to 70 Ma is widely represented, with the 110-90 Ma group being particularly abundant, and 70-20 Ma radiometric dates are found only on the west coast and offshore islands, with the younger of these coming only from the north. There is thus an apparent westward migration of activity, towards the trench and opposite to that in northern Chile.

Geochemical analyses of more than 500 volcanic and plutonic rocks from the Antarctic Peninsula, drawn largely from the British Antarctic Survey's collection with the guidance of Rex's (1976) compendium of radiometric ages, have been reported by Saunders et al., (1982b) and Weaver et al, (1982). These include South Shetland Is lavas ranging in age from 132 to 20 Ma (Pankhurst et. al., 1980) to represent the western, trench side, plutonic and volcanic rocks from the Danco Coast to represent the central part of the Peninsula (West, 1974), and mainly older lavas and plutons from the Bowman, Foyn and Oscar II Coast areas of the east coast (Fleet, 1968; Marsh, 1968; Stubbs 1968). Saunders et al., (1980) have combined these data sets, and the remainder of this section is essentially a précis of their findings.

The majority of rocks analysed have a typical calc-alkaline chemistry, showing moderate to low Fe-enrichment, high K_2O, Rb, Ba, Sr and Th levels, high Rb/Sr and low K/Rb and Na/K ratios, and strongly resemble calc-alkaline rocks of the Andean margin of South America. The S. Shetland Is lavas show some calc-alkaline affinities, but have lower K_2O and Rb and lower K/Rb ratios. Figure 11, from Saunders et al. (1980), shows that SiO_2, K_2O, Th and Ba all have higher values in east coast rocks (as do Rb, Ce and La), but there

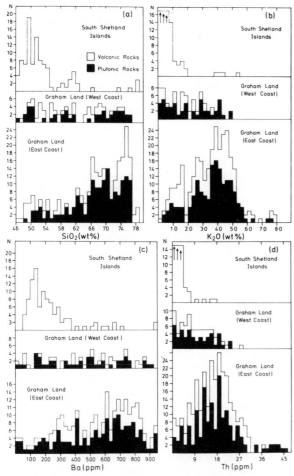

Fig. 11. Abundance of SiO_2, K_2O, Ba and Th in volcanic and plutonic rocks from the South Shetland Islands and the east and west coasts of the Antarctic Peninsula (from Saunders et al., 1980).

is in addition a steady increase in K, Rb, Th, Ce and La away from the trench at any particular silica concentration, indicating that the variations are not merely reflecting higher degrees of fractionation of east coast magmas. One interesting observation is that, in these respects, there is no significant difference between the data from plutonic and volcanic rocks, suggesting that both may represent true liquid compositions.

In contrast with the elements noted above, Sr, Ni, Cr and Zn decrease with increasing SiO_2, and Ba, Nb, Y, Zr and the heavy REE increase initially but then either level out or decrease with increasing SiO_2. Lest these results are being affected by the mineralogy of the fractionating phase, a real difficulty over such a wide compositional range, Saunders and others also plot several elements which they consider to have low bulk distribution co-

efficients over the entire range of compositions encountered. This produces clear linear relationships between K_2O and Rb, and between Rb and Th, with less scatter than on the silica plots, but the plot of Th against Nb is far from linear, and indicates that many elements which have quite low bulk distribution co-efficients in ocean ridge or continental rift systems (including Zr in particular) do not behave incompatibly in calc-alkaline rocks and cannot therefore be used as fractionation indices.

A further interesting result concerns the time- and space-dependence of the igneous geo-chemistry. Fortunately, activity on the Antarctic Peninsula is not totally correlated in time and space, so that some separation is possible. In particular, the South Shetland Is lavas, ranging in age from the earliest Cretaceous to the present day, are low in K, Rb, Th and La compared with lavas erupted on the east coast of the Peninsula, although the data have considerable scatter. Saunders and others conclude that compositional variations are therefore mainly space-dependent rather than time-dependent. It is perhaps surprising that this conclusion should emerge clearly from a subducting margin in which activity migrates towards the trench with time. To the extent that later plutonism in this situation creates rather than reworks continental crust, it displaces the trench from the older plutons, thereby flattening such spatial variation as exists while preserving any time variation. If spatial variation always dominates time variation, that dominance should be much more apparent where migration is away from the trench.

Saunders and others (1980) present an intere-sting discussion of the relevance of the Antarctic Peninsula results to problems of magma genesis. In particular they contrast behaviour in non-orogenic igneous suites, where the large-ion lithophile elements (K, Rb, Ba, Sr, Th, U, Pb) have bulk distribution co-efficients comparable with those of the high field strength (HFS) elements (Ti, P, Zr, Hf, Nb, Ta and the REE), with that in the calc-alkaline magma series, where the HFS elements show quite high bulk distribution coefficients at higher silica values. They argue that major-mineral phase fractionation in siliceous or hydrous conditions might contribute to this effect, but that its influence is unlikely to be dominant. While admitting also the possible significance of minor-element phase fractiona-tion and zone refining in the mantle above the Benioff zone, they point out that, for these to have any effect on the composition of the final melt, it is necessary also for a form of LIL/HFS decoupling to occur in the subducted oceanic crust and that, if this does occur, then these other mechanisms become of secondary importance.

The process in question is dewatering of the

subducted oceanic slab, which is not new as an explanation of calc-alkaline magma genesis by the volatile-induced partial melting of the over-lying mantle (Ringwood, 1974, 1977; Thorpe et al., 1976) and has many advantages over models involving the partial melting of the subducted slab itself. An important component of the argument of Saunders and others, however, is a recognition of the role of hydrothermal activity and halmyrolysis in increasing K, Rb, Sr, U, the $^{87}Sr/^{86}Sr$ ratio and possibly light REE in the upper part of the oceanic crust, in low-energy sites in alteration minerals (see, for example, Hart et al., 1974), within a relatively short time of formation at the ridge crest. They suggest that during dewatering in the subduc-tion zone these elements and other water-soluble components (including up to 10% dissolved silica) will be removed also, while HFS elements will remain behind unless the subducting slab itself starts to melt. This provides a simple mechanism for LIL/HFS decoupling, and the observed decoupling in turn supports the thesis that significant partial melting of the slab does not occur.

Even though hydration of oceanic crust near to the ridge crest may be essentially a steady process, there is scope for variability resulting from differences in sediment cover and the structural dependence on spreading rate. The supply of water from the subducted slab, and perhaps its mineral content, will in detail depend also on its age and on the subduction rate. Whatever set of processes within the mantle overlying the Benioff zone produce the final melt, the outcome of any particular stage will depend also upon the initial chemical and thermal conditions, which will be a function of the previous history of subduction. These are not aspects of the problem upon which Saunders and others are able to throw much light, beyond speculating that zone refining and mantle heterogeneity may produce the observed K-h variations, and mineral phase fractionation at high levels the observed Ni, Cr and Sr depletion and light REE enrichment. In particular, they point out that crustal fusion or contamination is in no way ruled out by the dehydration model.

Despite the reconnaissance nature of the radiometric age coverage of the Peninsula, the influence of one of the variables mentioned above, the age of the subducted slab, may already have been detected. Although the age of the ridge crest-trench encounter which ended subduction ranged from 50 Ma at the base of the Peninsula to 4 Ma off Brabant I, arc magmatic activity virtually ceased up to 50 Ma earlier in the south. Barker (1982) suggests this resulted from the very young age of the oceanic lithosphere being subducted there, which had all been produced at the distant but very rapidly-spreading Pacific-Phoenix plate boundary (Weissel et al. 1977;

Cande et al. 1982). The mechanism involved (DeLong et al., 1978) is the expulsion of bound water at shallower depth in the sub-duction zone, because of the greater heat content and thinner, less continuous sediment cover of young oceanic lithosphere. A pre-cursor of this effect, also noted along the Peninsula, might be the migration of magmatic activity towards the trench.

C3. Cordilleran Deformation - southernmost Andes

There is considerable variation in the style of so-called 'Andean', that is to say late Mesozoic and Cenozoic, deformation in the Scotia Arc region as a whole. Deformation in the northern Antarctic Peninsula, like that in the Central Andes, seems to be confined largely to open folding and block faulting. The rocks in the southernmost Andes, on the other hand, are highly deformed in a fashion comparable to those of the Pennine zone of the western Alps (see Nelson et al, 1980). This zone of intense deformation coincides with the extent of the Early Cretaceous marginal basin and seems to be associated with what can be described as an 'arc-continent collision' when the basin was uplifted and destroyed. Given the comparatively well-established tectonic setting of the area, and the excellent exposure, this makes the southernmost Andes an outstanding laboratory for the study of orogenic processes.

Detailed structural studies in the region have been underway for several years now, but the area to be covered is large and the work will take considerable time. That published so far includes studies in South Georgia on the marginal basin terrain (Dalziel, et al., 1975; Tanner, 1982) and along the rear (i.e. rifted continental margin or remnant arc) side of the marginal basin (Bruhn and Dalziel, 1977; Dalziel and Palmer, 1979; Bruhn, 1979; Nelson et al, 1980, Nelson, 1982). Here we summarize the state of knowledge concerning deformation of the Cordillera as it bears on the tectonic processes involved in form-ing a continental margin orogen.

Apart from the arc-trench gap terrain along the Pacific margin, which was presumably under-going compressive deformation but is now missing, the southernmost part of South America was undergoing extensional tectonics related to the development of the marginal basin in latest Jurassic and earliest Cretaceous time. At some time between early Albian and Coniacian or Turonian, the tectonic regime changed from extensional to compressional and the magmatic arc and marginal basin terrains appear to have been uplifted with respect to the stable part of the continent. In addition to uplift and compression, there is some evi-dence that a strike-slip component of the deformation was also initiated at this time.

The reason for the change from extension to compression is not certain, but it does coincide as closely as can be determined with a mid-Cretaceous world-wide increase in sea floor spreading rates and change in the SAM-AFR relative rotation pole (Dalziel, et al, 1974; Dalziel, 1981; Rabinowitz and LaBrecque, 1979). Recent indications of the sensitivity of back-arc extension to absolute plate motions (Chase, 1978b, Barker and Hill, 1980) underline the significance of this coincidence.

The intensity of deformation within the Cordillera increases from the magmatic arc (batholith) terrain on the Pacific side through the marginal basin terrain to reach a maximum in the supposed remnant arc terrain along the rifted continental margin (Figure 12). Compressional uplift (i.e. reverse faulting)of the marginal basin floor with respect to the continent appears to have been concentrated within the zone primarily affected by the earlier normal faulting which first created the basin.

There is excellent independent evidence that this penetrative deformation within the Cordillera was related to some physiographic uplift, in other words the initiation of the Andean proto-Cordillera, and the concomitant downwarp of the former rifted continental margin to create the so-called Magallanes basin, a classical foredeep, on the Atlantic side of the cordillera (Dalziel, and Palmer 1979). Eventually, through undetermined means, this led during the Cenozoic to the development of a foreland fold and thrust belt deforming the sedimentary infill of the Magallanes basin (see Section B4) and to final uplift of the Cordillera.

The deformation in the magmatic arc terrain was not extensive. The upper Jurassic volcanic rocks there are gently folded and quite strongly cleaved (see Dalziel and Cortes, 1972), and post-Jurassic structures affecting the earlier basement amount only to open warps. Within the marginal basin the rocks of the sedimentary infill are deformed mainly by tight asymmetric folds, verging towards the continent, with associated continentward-directed thrusts (Dalziel, et al., 1975; Bruhn and Dalziel, 1977). Folds and possible thrusts with opposing vergence occur in the infill on the continental (northern) side of South Georgia (Dalziel et al., 1975; Bruhn and Dalziel, 1977; Tanner, 1982b) and in southern Cordillera Darwin (Nelson et al, 1980).

The main enigma of the marginal basin terrain is why the infill is highly deformed and the (uplifted) ophiolitic rocks are deformed only along narrow shear zones. Obvious explanations lie in basement/cover décollement together with intense shortening across narrow shear zones and/or underthrusting (i.e. subduction on a small scale) of the denser marginal basin

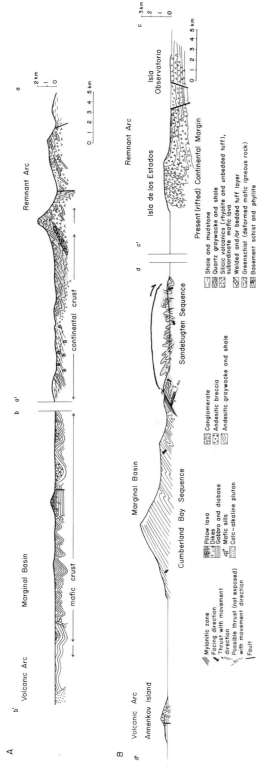

Fig. 12. Cross-sections of Andean Cordillera. For location see Fig. 3 (from Bruhn and Dalziel, 1977, with modification of section through South Georgia after Tanner, 1982).

floor. There is no independent evidence of a 'flipped' subduction zone. Until the deformation of the basin infill and the nature of deformation in the marginal basin as a whole are better understood the original width of the basin will remain uncertain.

As previously mentioned, the most intense deformation was concentrated along the rear wall of the marginal basin, i.e. along the rifted continental margin. The strains here are complex and polyphase, but nonetheless seem all to pre-date intrusion of 80-90 m.y. old granitic plutons associated (it is assumed) with migration of the magmatic arc after uplift and destruction of the marginal basin (Dalziel and Palmer, 1979). In both areas studied in detail to date (Isla de los Estados (Dalziel and Palmer, 1979) and Cordillera Darwin (Nelson et al, 1980)) the strains seem to result in shortening at right angles to the Pacific margin, and also in tectonic thickening of the rifted continental edge of the marginal basin (Figure 12). Taken with the independent evidence of the mid-Cretaceous (Cenomanian) onset of turbidite deposition in the contemporaneously down-warped foredeep, the Magallanes basin (Scott, 1966; Natland et al., 1974), this suggests a genetic relationship between penetrative deformation in the cordilleran terrain, initiation of uplift of the Cordillera, and downwarp of the foredeep (Dalziel and Palmer, 1979).

The deepest part of this intensely deformed terrain is exposed in a structural culmination of the pre-Jurassic basement in Cordillera Darwin (Nelson et al, 1980; Nelson, 1982) Here mid-Cretaceous (early Andean) deformation of the Upper Jurassic and Lower Cretaceous sedimentary cover has resulted in the near obliteration of the sub-Jurassic unconformity and the complete structural reworking of the Gondwanide basement by Andean structures, so that it became 'reactivated' in the same sense as the Hercynian basement in the Pennine zone of the western Alps. Cordillera Darwin is also the area of highest (amphibolite) grade regional metamorphism in the southern Andes. In this respect, therefore, it is a 'meta-morphic core complex' in the sense currently in vogue with workers in the North American Cordillera (see, for example, Davis and Coney, 1979).

It is interesting to note that the deformational fabrics of Cordillera Darwin are here interpreted as being compressional and appear to post-date any extensional tectonics.

Finally, it should be noted that the foreland fold and thrust belt on the Atlantic side of the Cordillera can be followed along strike to the east where the possibly oceanic floor of the eastern Falkland trough appears to be underthrust beneath the north Scotia Ridge. This variety of expression makes the area of great potential value for the study of such foreland deformational belts.

Acknowledgements We are grateful to Ramon Cabré, Chuck Drake, Oscar Gonzalez-Ferran, Eric Nelson and Andrew Saunders for constructive criticism, and to Janet Thomson for extensive contributions to Figure 4. Preparation of this Report was supported by NERC research grant GR3/3279 (PFB) and NSF grant DPP 7820629 (IWDD).

References and Geodynamics Bibliography

About one quarter of the bibliography consists of publications not referred to in the text. In addition, the following books and periodicals contain many other papers of broad interest in Scotia Arc studies.

Adie, R.J. 1972. Antarctic Geology and Geophysics. Universitetsforlaget, Oslo.
Craddock, C. 1982. Antarctic Geoscience. Univ. Wisconsin Press. Madison.
Gonzalez-Ferran, O. 1976. Andean and Antarctic Volcanology Problems. IAVCEI Special Series. Rome.
Antarctic Journal of the United States.
British Antarctic Survey Bulletin and Scientific Reports.

Adie, R.J. 1964. Geological History. In: Priestley, R.E., Adie, R.J. and G. de Q. Robin, eds., Antarctic Research, pp. 118-162, Butterworth, London.
Adie, R.J. 1972. Evolution of volcanism in the Antarctic Peninsula. in Adie, R.J. ed. Antarctic Geology and Geophysics. Universitetsforlaget, Oslo, 137-141.
Alarcon, B., Ambrus, J., Olcay, and Vieira, C. 1976. Geologia del Estrecho Gerlache entre los paralelos 64° Lat. Sur, Antarctica Chilena. Ser. Cient. Inst. Antar. Chileno 4, 7-51.
Allen, A., 1966. Seismic refraction investigations in the Scotia Sea. Br. Antarct. Surv. Sci. Rep. 55, 44pp.
Ashcroft, W.A. 1972. Crustal Structure of the South Shetland Islands and Bransfield Strait. Br. Antarct. Surv. Sci. Rep. 66, 43pp.
Baker, P.E. 1968. Comparative volcanology and petrology of the Atlantic island-arcs. Bull. Volcan. 32, 189-206 .
Baker, P.E., 1976. Volcanism and plate tectonics in the Antarctic Peninsula and Scotia Arc. In Gonzalez-Ferran, O., ed. Proceedings of the Symposium on Andean and Antarctic Volcanology Problems. Rome, IAVCEI, 347-356.
Baker, P.E. 1978. The South Sandwich Islands: II Petrology of the volcanic rocks. Br. Antarct. Surv. Sci. Rep., 93, 34pp.
Baker, P.E., Gonzalez-Ferran, O. and Vergara, M., 1976. Geology and geochemistry of Paulet Island and the James Ross Island Volcanic Group, In. Gonzalez-Ferran, O., ed. Procee-

dings of the Symposium on Andean and Antarctic Volcanology Problems, Rome, IAVCEI, 39-47.

Baker, P.E., McReath, I., Harvey, M.R., Roobol, M.J. and Davies, T.G., 1975. The geology of the South Shetland Islands: V. Volcanic Evolution of Deception Island. Br. Antarct. Surv. Sci. Rep. No. 78 81pp.

Baker, P.E., Buckley, F. and Rex, D.C., 1977. Cenozoic volcanism in the Antarctic. Phil. Trans. Roy. Soc. Lond. Ser. B, 279. 131-142.

Barazangi, M., Pennington, W. and Isacks, B. 1975. Global study of seismic wave attenuation in the upper mantle behind island arcs using pP waves. J. geophys. Res., 80, 1079-1092.

Barker, P.F., 1970. Plate tectonics of the Scotia Sea region. Nature, London, 228, 1293-1297.

Barker, P.F., 1972a. Magnetic lineations in the Scotia Sea. In: Adie, R.J. ed Antarctic Geology and Geophysics, Universitetsforlaget Oslo, 17-26.

Barker, P.F. 1972b. A spreading centre in the east Scotia Sea. Earth planet. Sci. Lett., 15, 123-132.

Barker, P.F. 1976. The tectonic framework of Cenozoic volcanism in the Scotia Sea region, A review. In Gonzalez-Ferran, O, ed. Symposium on Andean and Antarctic volcanology Problems. Proc. Santiago, Chile, IAVCEI, Spc. Ser. 330-346.

Barker, P.F. 1982. The Cenozoic subduction history of the Pacific margin of the Antarctic Peninsula: ridge crest-trench interactions. J. geol. Soc. Lond. 139. 787-801.

Barker, P.F. and Burrell, J., 1977. The opening of Drake Passage, Marine Geol, 25, 15-34.

Barker, P.F. and J. Burrell, 1982. The influence upon Southern Ocean circulation, sedimentation and climate of the opening of Drake Passage. In Craddock, C. ed. Antarctic Geoscience, Madison, Univ. Wisconsin Press, 377-385.

Barker, P.F. and D.H. Griffiths, 1972. The evolution of the Scotia Ridge and Scotia Sea. Phil. Trans. Roy. Soc. Lond. Ser. A, 271 151-183.

Barker, P.F. and Griffiths, D.H. 1977. Towards a more certain reconstruction of Gondwanaland. Phil. Trans. Roy. Soc. Lond. Ser. B, 279. 143-159.

Barker, P.F. and Hill, I.A. 1980. Asymmetric spreading in back-arc basins. Nature, 285, 652-4.

Barker, P.F. and Hill, I.A. 1981. Back-arc extension in the Scotia Sea. Phil. Trans. Roy. Soc. Lond. Ser. A, 300, 249-262.

Barker, P.F. and Jahn, R.A. 1980. A marine geophysical reconnaissance of the Weddell Sea. Geophys. J.R. astr. Soc. 63, 271-283.

Barker, P.F. Dalziel, I.W.D. and others 1976 Initial Reports of the Deep Sea drilling Project Vol. 36, Washington, U.S. Government Printing Office, 1080pp.

Barker, P.F., Hill, I.A. Weaver, S.D. and

Pankhurst, R.J. 1982. The origin of the eastern South Scotia Ridge as an intra-oceanic island arc. In Craddock, C. ed. Antarctic Geosciences, Madison, Univ. Wisconsin Press, 203-211.

Barron, E.J. Harrison, C.G.A. and Hay, W.W. 1978. A revised reconstruction of the Southern Continents. EOS, 59, 436-449.

Barton, C.M., 1965. The geology of the South Shetland Islands. III. The stratigraphy of King George Island. Br. Antarct. Surv. Sci. Rep. No. 44, 33pp.

Bell, C.M. 1973. The geology of southern Alexander Island: Br. Antarct. Surv. Bull. 33 and 34, 1-16.

Bell, C.M., 1974. Geological observations in northern Alexander Island: Br. Antarct. Surv. Bull. 39, 35-44.

Bell, C.M., 1975. Structural geology of parts of Alexander Island: Br. Antarct. Surv. Bull. 41 and 42, 43-58.

Bell C.M., Mair, B.F. and Storey, B.C., 1977. The geology of part of an island-arc marginal basin system in southern South Georgia, Br. Antarc. Surv. Bull. 46, 109.

Bergh, H.W. 1977. Mesozoic sea floor off Dronning Maud Land, Antarctica. Nature 267, 686-7.

Bergh, H.W. and Norton, I.O. 1976. Prince Edward fracture zone and the evolution of the Mozambique Basin. J. geophys. Res., 81 5221-39.

Brett, C.P. 1977. Seismicity of the South Sandwich Islands region. Geophys. J.R. astr. Soc. 51, 453-464.

Brett, C.P. and Griffiths, D.H. 1975. Seismic wave attenuation and velocity anomalies in the eastern Scotia Sea. Nature, 253, 613-4.

Bruhn, R.L., 1979. Rock structures formed during back-arc basin deformation in the Andes of Tierra del Fuego: Geol. Soc. Amer. Bull. 90, 998-1012.

Bruhn, R.L. and Dalziel, I.W.D. 1977. Destruction of the early Cretaceous marginal basin in the Andes of Tierra del Fuego. In: Talwani, M. and W.C. Pitman II (eds.) Island Arcs, Deep Sea Trenches and Back-Arc Basins. A.G.U. Washington D.C. 395-406.

Bruhn, R.L. Stern, C.R. and DeWit, M.J. 1978. Field and Geochemical Data Bearing on the Development of aMesozoic Volcano-Tectonic Rift Zone and Back-Arc Basin in Southernmost South America. Earth Planet. Sci. Lett, 41 32-46.

Cande, S.C., Herron, E.M. and Hall B.R., 1982. The Early Cenozoic tectonic history of the southeast Pacific. Earth planet. Sci. Lett. 57, 63-74.

Cecioni, G. 1955. Noticias preliminaries sobre el hallazgo del Paleozoico superior en el Archipielago Patagonico. Boln Inst. Geol. Fac. Cienc. Fis y Matem. Univ. Chile, 6, 241-55.

Chase, C.G. 1978a. Plate kinematics: the

Americas, East Africa, and the rest of the world. Earth planet. Sci. Lett, 37. 355-368.

Chase, C.G. 1978b. Extension behind island-arcs and motions relative to hot spots. J. geophys. Res. 83, 5385-8.

Cingolani, C.A. and Varela, R. 1976. Investigationes Geologicas y Geochronologicas en al extremo sur de la Isla Gran Malvina, sector de Cabo Belgrano (Cape Meredith), Islas Malvinas.
Actas del Sexto Congreso Geologico Argentino.

Clarkson, P.D. and Brook, M. 1977. Age and positions of the Ellsworth Mountains crustal fragment, Antarctica. Nature, 265, 615-16.

Craddock, C. and Hollister, C.D. 1976. Geologic evolution of the southeast Pacific basin. In Hollister, C.D., Craddock, C. and others. Initial reports of the Deep Sea Drilling Project. Vol. 35. Washington, D.C., U.S. Government Printing Office, 723-743.

Dalziel, I.W.D. 1972a. Large scale folding in the Scotia Arc. In Adie, R.J. ed. Antarctic Geology and Geophysics, Universitetsforlaget, Oslo, 47-55

Dalziel, I.W.D. 1972b. K-Ar dating of rocks from Elephant Island, South Scotia Ridge, Geol. Soc. Amer. Bull. 83, 1887-1894.

Dalziel, I.W.D. 1974. Evolution of the Margins of the Scotia Sea, in Burk, C.A. and Drake, C.L. eds., The geology of continental margins: New York, Springer-Verlag New York, Inc., 567-579.

Dalziel, I.W.D. 1981. Back-arc extension in the Southern Andes: a review and critical reappraisal. Phil. Trans. Roy. Soc. Lond. A, 300, 300-335.

Dalziel, I.W.D. 1982. The Pre-Middle Jurassic history of the Scotia Arc: a review and progress report. In Craddock, C. ed., Antarctic Geosciences, Madison, Univ. Wisconsin Press, 111-126.

Dalziel, I.W.D. and Cortes, R. 1972. Tectonic style of the southernmost Andes and the Antarctandes, 24th Int. geol. Congress. Montreal, Sect. 3, 316-27

Dalziel, I.W.D. and Elliot, D.H. 1971. The evolution of the Scotia Arc: Nature, 233 246-252.

Dalziel, I.W.D. and Elliott, D.H. 1973. The Scotia Arc and Antarctic margin., In: Nairn A.E.M. and Stehli, F.G. eds. The Ocean Basins and Margins Vol. 1. The South Atlantic. Plenum Press, 171-246.

Dalziel, I.W.D. and Palmer, K.F. 1979. Progressive deformation and orogenic uplift at the southern-most extremity of the Andes. Geol. Soc. Amer. Bull. 90, 259-280.

Dalziel, I.W.D. Caminos, R., Palmer, K.F., Nullo, F. and Casanova, R. 1974. South extremity of the Andes: geology of Isla de los Estados, Argentine Tierra del Fuego. Bull. Am. Ass. Petrol. Geol. 58. 2502-12.

Dalziel, I.W.D. DeWit, M.S. and Palmer, K.F.

1974. A fossil marginal basin in the southern Andes. Nature, 250, 291-294.

Dalziel, I.W.D. Dott, R.H. Jr., Winn, R.D. Jr. and Bruhn, R.L. 1975. Tectonic relationships of South Georgia Island to the southernmost Andes, Geol. Soc. Amer. Bull., 86, 1034-1040.

Dalziel, I.W.D., Kligfield R., Lowrie, W. and Opdyke, N.D. 1973. Palaeomagnetic data from the southernmost Andes and Antarctandes. In Tarling, D.H. and Runcorn, S.K. eds. Implications of Continental Drift to the Earth Sciences, New York, Academic Press, 87-101.

Davey, F.J. 1972a. Gravity measurements over Burdwood Bank, Mar. Geophys. Res. 1, 428-435.

Davey, F.J. 1972b. Marine gravity measurements in Bransfield Strait and adjacent areas, In Adie, R.J. ed. Antarctic geology and geophysics. Oslo, Universitetsforlaget, 39-46.

Davis, G.H. and Coney, P.J. 1979. Geologic development of the Cordilleran metamorphic core complexes. Geology 7, 120-4.

DeLong, S.E., Fox, P. and McDowell, F.W. 1978. Subduction of the Kula Ridge at the Aleutian trench. Geol. Soc. Amer. Bull. 86, 1034-1040.

Del Valle, R, Morelli, J. and Rinaldi, C, 1974. Manifestacions cuproplumbifera 'Don Bernabe' Isla Livingston, Islas Shetland del Sur, Antatida Argentina. Contr. Inst. Antar. Argentino, 175 3-35.

DeWit, M.J. 1977. The evolution of the Scotia Arc as the key to the reconstruction of south-west Gondwanaland, Tectonophysics, 37, 53-81.

DeWit, M.J. and Stern, C.R. 1976. A model for ocean-floor metamorphism, seismic layering and magnetism, Nature, 264, 615-619.

DeWit, M.J. and Stern, C.R. 1978. Pillow Talk, Jour. Volcanology and Geothermal Res. 4, 1/2, 55-80.

DeWit, M.J., Dutch, S., Kligfield, R., Allen, R. and Stern, C.R. 1977. Deformation, serpentinization and emplacement of a dunite complex, Gibbs Island, South Shetland Islands-possible fracture zone tectonics. Journal of Geology, 85, 745-762.

Dietz, R.S. and Holden, J.C. 1970. Reconstruction of Pangaea: breakup and dispersion of continents Permian to Present, J. geophys. Res. 75, 4939-4956.

Dott, R.H. 1974. Palaeocurrent analysis of severely deformed flysch-type strata- A case study from South Georgia Island, Jour. Sed. Petrology, 44, 1166-1173.

Dott, R.H. Jr. 1976. Contrasts in Tectonic History along the Eastern Pacific Rim. In The Geophysics of the Pacific Ocean Basin and its Margin, Geophysical Monograph 19 Washington, D.C. Am. Geophys. Union, 299-308.

Dott, R.H. Jr. Winn, R.D., DeWit., M.J. and Bruhn, R.L. 1977. Tectonic and sedimentary significance of Cretaceous Tekenika Beds of Tierra del Fuego: Nature, 266, 620-622.

Dott, R.H. Jr., Winn, R.L. Jr. and Smith, C.H.L. 1982. Relationships of Late Mesozoic and Early

Cenozoic Sedimentation to the Tectonic Evolution of the Southernmost Andes and Scotia Sea, in Craddock, C. ed. Antarctic Geoscience, Madison Univ. Wisconsin Press, 193-202.

Edwards, C.W. 1982. Further palaeontological evidence of Triassic sedimentation in Western Antarctica, In Craddock, C. ed. Antarctic Geoscience, Madison, Univ. Wisconsin Press, 325-320.

Edwards, C.W. 1980. Early Mesozoic marine fossils from central Alexander Island, Br. Antarct. Surv. Bull. 49. 33-58.

Elliott, D.H. 1972. Antarctic geology and drift reconstructions, In Adie, R.J. ed. Antarctic geology and geophysics, Oslo, Universitetsforlaget 849-858.

Elliott, D.H. 1975. Tectonics of Antarctica - A review, Am. Jnl. Sci. 275-A, 45-106.

Elthon, D. 1979. High magnesia liquids as the parental magma for ocean floor basalts. Nature. 278, 514-8

Elthon, D. and Stern, C, 1978. Metamorphic petrology of the Sarmiento ophiolite complex, Chile: Geology, 6, 464-468.

Ewing, J.I.,Ludwig,W.J.,Ewing, M. and Eittreim, S.L. 1971. Structure of the Scotia Sea and Falkland Plateau. J. geophys. Res. 76, 7118 - 7137.

Farrar, E.,Clark, A.H., Haynes, S.J.,Quirt, G.S., Conn, H. and Zentilli, M. 1970. K-Ar evidence for the post-Palaeozoic migration of granitic intrusion foci in the Andes of northern Chile, Earth planet, Sci. Lett. 10, 60-66.

Farrar, E., McBride , S.L. and Rowley, P.D. 1982. Ages and tectonic implications of Andean plutonism in the southern Antarctic Peninsula, In. Craddock, C. ed. Antarctic Geoscience, Madison, Univ. Wisconsin Press, 349-355.

Fleet, M. 1968. The geology of the Oscar II Coast, Graham Land, Br. Antarct. Surv. Sci. Rep. 59, 46pp.

Forsyth, D.W. 1975. Fault plane solutions and Tectonics of the South Atlantic and Scotia Sea, J. geophys. Res., 80, 1429-1443.

Forsyth, D.W. and Uyeda, S. 1975. On the relative importance of the driving forces of plate motion, Geophys. J.R. astr. Soc. 43, 163-200

Forsythe, R. 1982. The late Palaeozoic to early Mesozoic evolution of southern South America, a plate tectonic interpretation. J. geol. Soc. Lond. 139, 671-682.

Frankel, A. and McCann, W. 1979. Moderate and large earthquakes in the South Sandwich arc: indicators of tectonic variation along a sub-duction zone. J. gephys. Res., 84, 5571-7.

Fuenzalida, R.P. 1976. The Magellan fault zone, In. Gonzalez-Ferran, O. ed. Andean and Antarctic Volcanology Problems, Rome, IAVCEI 373-391.

Gledhill, A. and Baker, P.E. 1973. Strontium isotope ratios in volcanic rocks from the South Sandwich Islands, Earth planet. Sci. Lett., 19, 369-372.

Gledhill, A., Rex, D.C. and Tanner, P.W.G. 1982.

Rb-Sr and K-Ar geochronology of rocks from the Antarctic Peninsula between Anvers Island and Marguerite Bay, IN Craddock, C. ed. Antarctic Geoscience, Madison, Univ. Wisconsin Press. 315-323.

Gonzalez-Ferran, O. 1972. Distribution, migration, and tectonic control of Upper Cenozoic volcanism in West Antarctica and South America. In Antarctic Geology and Geophysics, ed. R.J. Adie, Universitetsforlaget, Oslo 173-180.

Gonzalez-Ferran, O. and Baker, P.E. 1974. Isla Paulet y el Volcanismo Reciente en las Islas del Weddell Noroccidental, Antarctica, Rev. Geograf. de Chile 69. 25-43.

Gonzalez-Ferran, O. and Katsui, Y. 1970. Estudio Integral del volcanismo cenozoico superior de las Islas Shetland del Sur, Antarctica, Inst. Ant. Ch. Ser. Cient. 1, (2) 123-174.

Gonzalez-Ferran, O. and Vergara, M. 1972. Post-Miocene volcanic petrographic provinces of West Antarctica and their relation to the Southern Andes of South America, In Antarctic Geology and Geophysics, ed. R.J. Adie, Universitetsforlaget, Oslo, 187-195.

Griffiths, D.H. and Barker, P.F. 1972. Review of marine geophysical investigations in the Scotia Sea, In Antarctic Geology and Geophysics, ed. R.J. Adie, Universitetsforlaget, Oslo 3-12.

Grikurov, G.E. 1972. Tectonics of the Antarct-andes, In Antarctic Geology and Geophysics, ed. R.J. Adie, Universitetsforlaget, Oslo, 163-168.

Grikurov, G.E. 1972. Geologiya Antarkticheskogo polvostrova (Geology of the Antarctic Peninsula). Moscow, Nauka Publishers, Academy of Sciences of the U.S.S.R., 119pp. (English translation: New Delhi, Amerind Pub. Co. Pvt. Ltd. for Nat. Sci. Found).

Grikurov, G.E. and Lopatin, B.G. 1975. Structure and evolution of the West Antarctic part of the circum-Pacific mobile belt, In Campbell, K.S.W. ed. Gondwana geology, papers presented at the third Gondwana symposium, Canberra. Australia National Univ. Press. 639-650.

Grikurov, G.E., Krylov, A.Ya., Polyakov, M.M. and Tsovbun, Ya. N. 1970. Age of rocks in the northern part of the Antarctic Peninsula and on the South Shetland Islands (According to Potassium-argon data). Soviet Antarctic Expedition Information Bulletin, 8, 61-63.

Halpern, M. 1971. Evidence for Gondwanaland from a review of West Antarctic radiometric ages. (In Quam, L.O., ed. Research in the Antarctic, Am. Assoc. Adv. Sci, 171-30

Halpern, M. 1973. Regional geochronology of Chile south of 50° latitude, Geol. Soc. Amer. Bull. 84, 2407-22

Halpern, M. and Fuenzalida, R. 1978. Rubidium-strontium geochronology of a transect of the Chilean Andes between latitude 45° and 46°S, Earth planet. Sci. Lett., 41, 60-66.

Halpern, M. and Rex, D.C. 1972. Time of folding of the Yahgan Formation and age of the Tekenika Beds, southern Chile, South America,

Geol. Soc. Amer. Bull, 83, No. 6 1881-86.

Halpern, M, Stipanicic and Toubes, R.O. 1975. Geochronologia (Rb/Sr) en los Andes Australes Argentinos, Rev. Asoc. Geol. Argentina, 30, 180-192.

Hamer, R.D. and Moyes, A.B. 1982. Composition and origin of garnet from the Antarctic Peninsula Volcanic Group of Trinity Peninsula. J. geol. Soc. Lond. 139, 713-720.

Harrington, P.K., Barker, P.F. and Griffiths, D.H. 1972. Crustal structure of the South Orkney Islands area from seismic refraction and magnetic measurements. In: Adie, R.J. ed., Antarctic Geology and Geophysics, Universitetsforlaget, Oslo, 27-32.

Hart, S.R., Erlank, A.J. and Kable, E.J.D. 1974. Sea floor basalt alteration: some chemical and strontium isotopic effects. Contrib. Mineral. Petrol. 44, 219-230.

Hawkes, D.D. 1961. The Geology of the South Shetland Islands, II. The Geology and Petrology of Deception Island. Falkland Islands Dependencies Survey Scientific Reports 27. 43pp.

Hawkes, D.D. 1962. The structure of the Scotia Arc. Geol. Mag., 99, 85-91.

Hawkes, D.D. 1981. Tectonic segmentation of the northern Antarctic Peninsula. Geology, 9, 220-4.

Hawkes, D.D. 1982. Metalliferous mineralisation in the northern Antarctic Peninsula. J. geol. Soc. Lond. 139, 803-9.

Hawkesworth, C.J., O'Nions, R.K., Pankhurst, R.J., Hamilton, P.J. and Evenson, N.M. 1977. A Geochemical study of Island-arc and back-arc tholeiites from the Scotia Sea. Earth planet. Sci. Lett, 36, 253.

Hayes, D.E. 1974. Continental margin of western South America. In. Burk, C.A. and Drake, C.L. eds. The Geology of Continental Margins. New York. Springer-Verlag, 581-590.

Hayes, D.E. and Ewing, M. 1970. Pacific Boundary Structure. In The Sea, 4 Part II, ed. A.E. Maxwell, Wiley-Interscience, New York .

Herron, E.M. 1971.. Crustal plates and sea floor spreading in the southeastern Pacific. Antarctic Research Series, 15, 229-237. Am. Geophys. Union. Washington D.C.

Herron, E.M. and Hayes, D.E. 1969. A geophysical study of the Chile Ridge, Earth planet. Sci. Lett, 6, 77-83.

Herron, E.M. and Tucholke, B.E., 1976. Sea-floor magnetic patterns and basement structure in the southeastern Pacific, in Hollister, C.D., Craddock, G, and others, Initial reports of the Deep Sea Drilling Project, 35, Washington, D.C. U.S. Govt. Printing Office, 263-278.

Herron, E.M., Bruhn, R, Winslow, M., and Chuaqui, L. 1977. Post Miocene Tectonics of the Margin of Southern Chile. In Talwani, M. and W.C. Pitman III. eds. Maurice Ewing Series, v.1: Island Arcs, Deep Sea Trenches and Back-arc Basins, Washington, D.C. Am. Geophy. Union 273-84.

Herron, E.M., Cande, S.C. and Hall, B.R. 1981. An active spreading center collides with a subduction zone: a geophysical survey of the Chile margin triple junction. Mem. geol. Soc. Amer. 154. 683-701

Hill, I.A. 1978. A model for the formation of the Central Scotia Sea. Abstr. UKGA2, Geophys. J.R. astr. Soc., 53, 164.

Hill, I.A. and Barker, P.F. 1980. Evidence for Miocene back-arc spreading in the central Scotia Sea. Geophys. J.R. astr. Soc., 63, 427-440.

Hollister, C.D., Craddock, C. and others. Initial Reports of the Deep Sea Drilling Project. 35 Washington, D.C. U.S. Govt. Printing Office. 930pp.

Hooper, P.R. 1962. The petrology of Anvers Island and adjacent islands. Scient. Rep. Falkd. Is. Depend. Surv. 34, 69pp.

Hyden, G. and Tanner, P.W.G. 1981. Late Palaeozoic - early Mesozoic fore-arc basin sedimentary rocks of the Pacific margin in West Antarctica. Geol. Rundschau, 70, 529-541.

Isacks, B. and Molnar, P. 1971. Distribution of stresses in the descending lithosphere from a global survey of focal mechanism solutions of mantle earthquakes. Rev. Geophys. Space Phys., 9, 103-174.

Jahn, R.A. 1978. A preliminary interpretation of Weddell Sea magnetic anomalies. Abstr. UKGA2. Geophys. J.R. astr. Soc., 53, 164.

Jakes P and Gill, J. 1970. Rare earth elements and the island-arc tholeiitic series, Earth Planet. Sci. Lett. 9, 17-28

Johnson, B.D., Powell, C.McA. and Veevers, J.H. 1982. Implications regarding the evolution of the Scotia Arc from a plate tectonic model of the dispersion of the Gondwanic continents. IN Craddock, C. ed. Antarctic Geoscience, Madison, Univ. Wisconsin Press, 418.

Kamenev, E.N. 1975. Skladchatiyeh osadochniyeh i intrvziniyeh obrazovaniya Berega Lassitera i Berega Bleka (Antarkticheskii Polovostrov) (Folded sedimentary and intrusive formations of the Lassiter Coast and the Black Coast and the Black Coast (Antarctic Peninsula)): 17th Soviet Antarct. Exped. Winter Res. 1971-1973 65, 215-222.

Karig, D.E. 1971. Origin and development of marginal basins in the Western Pacific, J. Geophys. Res. 76, 2542-2561.

Karig, D.E. 1972. Remnant arcs. Geol. Soc. Amer. Bull. 83, 1057-68.

Karig, D.E. 1977. Growth patterns on the upper trench slope. In Talwani, M. and Pitman, W.C. III Eds, Island Arcs, Deep Sea Trenches and Back-Arc Basins. Maurice Ewing Series I. Washington D.C. Am. Geophys. Union, 273-284.

Katz, H.R. 1972. Plate tectonics and orogenic belts in the south-east Pacific. Nature 237, 331-2.

Katz, H.R. 1973a. Contrasts in tectonic

evolution of orogenic belts in the south-east Pacific, J.R. Soc. N.Z., 3, 333-61.

Katz, H.R. 1973b. Time of folding of the Yahgan Formation and age of the Tekenika Beds, southern Chile, South America: discussion. Geol. Soc. Amer. Bull., 84, 1109-12.

Katz, H.R. and Watters, W.A. 1966. Geological investigation of the Yahgan Formation (Upper Mesozoic) and associated igneous rocks of Navarino Island, southern Chile, N.Z. Jl. Geophys. 9. 323-59.

Kellogg, K.S. and Reynolds, R.L. 1978. Palaeomagnetic results from the Lassiter Coast, Antarctic Peninsula. J. geophys. Res.83 2293-2300.

LaBrecque, J.L. and Barker, P.F. 1981. The age of the Weddell Basin, Nature , 290, 489-492.

LaBrecque J.L. and Rabinowitz, P.D. 1977. Magnetic anomalies bordering the continental margin of Argentina. Amer. Assoc. Petr. Geol. Map Series AAPG. Tulsa.

LaBrecque, J.L., Kent D.V. and Cande, S.C. 1977. Revised magnetic polarity time scale for late Cretaceous and Cenozoic time. Geology, 5, 330-335.

Larson, R.L. and Ladd, J.W. 1973. Evidence for the opening of the South Atlantic in the Early Cretaceous. Nature, 246, 210-212.

Laudon, T.S. 1972. Stratigraphy of eastern Ellsworth Land, in Adie, R.J. ed., Antarctic geology and geophysics (Oslo) Universitets-forlaget, 215-223.

Laudon, T.S. 1982. Igneous geochemistry of eastern Ellsworth Land. In. Craddock, C. ed. Antarctic Geoscience. Madison, Univ. Wisconsin Press, 775-785.

Lopatin, B.G., Krylov, A.Y. and O.A. Aliapyshev. 1974. Osnovniye tektonomagmatecheskiye etapi v rasvetee zemle Marie Byrd Berega Eightsa (Zapadnaya Antarctida) po radeogennim dannim (Major tectonomagnetic stages in the development of Marie Byrd Land and Eights Coast (West Antarctica) determined radiometrically): Antartika, Doklady Komissii, No. 13, 52-60.

Ludwig, W.J. and Rabinowitz, P.D. 1982. The collision complex of the North Scotia Ridge. J. geophys. Res. 87, 3731-40.

Ludwig, W.J. Windisch, C.C. Houtz, R.E. and Ewing, J.I. Structure of Falkland Plateau and offshore Tierra del Fuego. In Watkins, J.S. et al. eds. Geological and Geophysical Investigations of Continental Margins. Mem. 29. Amer. Assoc. Petr. Geol. Tulsa. 125-137.

Macdonald, K.C. and Holcombe, T.L. 1978. Inversion of magnetic anomalies and sea floor spreading in the Cayman Trough. Earth planet. Sci. Lett. 40. 407-414.

McNutt, R.H., Crocket, J.H.,Clark, A.H.,Caelles, J.C.,Farrar, E. and Haynes, S.J. 1975. Initial $^{87}Sr/^{86}Sr$ ratios of plutonic and volcanic rocks of the Central Andes between latitudes 26° and 29°South. Earth planet. Sci. Lett. 27, 305- 313.

Marsh, A.F. 1968. Geology of parts of the Oscar II and Foyn Coasts, Graham Land. Unpublished Ph.D. Thesis, University of Birmingham 291pp.

Miller, H. 1982. Geological comparison between the Antarctic Peninsula and southern South America. In. Craddock, C. ed Antarctic Geoscience. Madison. Univ. Wisconsin Press, 127-134.

Minster, J.B. and Jordan, T.H. 1978. Present-day plate motions. J. geophys. Res. 83, 5331-5354.

Molnar, P. and Atwater, T. 1978. Interarc spreading and Cordilleran tectonics as alternates related to the age of subducted oceanic lithosphere. Earth Planet. Sci. Lett. 41, 330-340.

Munizaga, F., Aguirre, L. and Herve, F. 1973. Rb/Sr ages of rocks from the Chilean metamorphic basement, Earth planet. Sci. Lett. 18, 87-92.

Murauchi, S., Asanuma, T. and Saki, K. 1976. Seafloor spreading in the Shikoku Basin, South of Japan. Abstr. with Progr. 1st. Ewing Symp., Harriman, New York.

Natland, M.L. Gonzalez, E. Canon. A and Ernst, M. 1974. A system of stages for correlation of Magallanes Basin sediments. Mem. geol. Soc. Am., No. 139. 126pp.

Nelson, E.P. 1982. Post-tectonic uplift of the Cordillera Darwin orogenic core complex: evidence from fission track geo chronology and closing temperature-time relationships. J. geol. Soc. Lond. 139, 755-761.

Nelson, E.P., Dalziel, I.W.D. and Milnes, A.G. 1980. Collision-style orogenesis in the southernmost Andes. Eclog. geol. Helv. 73 727-751.

Nelson, P.H.H. 1975. The James Ross Island Volcanic Group of North-East Graham Land. Br. Antarct. Surv. Sci. Repts. No., 54 62pp.

Norton I. 1982. Palaeomotion between Africa, South America and Antarctica and implications for the Antarctic Peninsula. In. Craddock, C. ed. Antarctic Geoscience. Madison. Univ. Wisconsin Press, 99-106.

Norton, I. and Sclater, J.G. 1979. A model for the evolution of the Indian Ocean and the break-up of Gondwanaland. J. geophys. Res., 84, 6803-6830.

Pankhurst, R.J. 1982a. Sr-isotopes and trace element geochemistry of Cenozoic volcanics from the Scotia Arc and the northern Antarctic Peninsula. In Craddock, C. ed. Antarctic Geoscience. Madison. Univ. Wisconsin, Press, 229-234.

Pankhurst, R.J. 1982b. Rb-Sr geochronology of Graham Land, Antarctica. J. geol. Soc. Lond. 139, 701-711.

Pankhurst, R.J., Weaver, S.D. Brook, M. and Saunders, A.D. 1980. K-Ar chronology of Byers Peninsula, Livingston Island, South Shetland Islands, Bull. Br. Antarct. Surv . 49, 277-82.

Pettigrew, T.H. 1981. The geology of Annenkov Island, Br. Antarct. Surv. Bull. 53, 312-254.

Rabinowitz, P.D. and LaBrecque, J.L. 1979. The Mesozoic South Atlantic ocean and evolution of its continental margins. J. geophys. Res. 84, 5973-6002.

Renner, R.G.B. Dikstra, B.J. and Martin, J.L. 1982. Aeromagnetic surveys over the Antarctic Peninsula. In. Craddock C. ed. Antarctic Geoscience. Madison, Univ. Wisconsin Press, 363-370.

Rex, D.C. 1972. K-Ar age determinations on volcanic and associated rocks from the Antarctic Peninsula and Dronning Maud Land. In Adie, R.J. ed. Antarctic geology and geophysics. Oslo, Universitetsforlaget 133-36.

Rex, D.C. 1976. Geochronology in relation to the stratigraphy of the Antarctic Peninsula. Br. Antarct. Surv. Bull. 43, 49-58.

Rex, D.C. and Baker, P.E. 1973. Age and petrology of the Cornwallis Island granodiorite. Br. Antarct. Surv. Bull., 32, 55-61.

Ringwood, A.E. 1974. Petrological evolution of island arc systems. J. Geol. Soc. London,130 183.

Ringwood, A.E. 1977. Petrogenesis in island arcs, In Island arcs, deep sea trenches and back-arc basins, Maurice Ewing Series I. ed. M. Talwani and W.C. Pitman III. American Geophysical Union, Washington, D.C. 311-324.

Rivano, S. and Cortes. R. 1975. Nota preliminar sobre el Hallazgo de rocas metamorphicas en la Isal Smith (Shetland del Sur Antarctica Chilena). Instituto Antarctico Chileno, Serie Cientifica, III, 9-14.

Roach, P.J. 1978. The nature of back-arc extension in Bransfield Strait. Geophys. J.R. Astron Soc. 53, 165. (asbtr. only).

Rowley, P.D. and Pride, D.E. 1982. Metallic mineral resources of the Antarctic Peninsula (review). In Craddock, C. ed. Antarctic Geoscience. Madison Univ. Wisconsin Press, 859-870.

Rowley, P.D. and Williams, P.L. 1982. Geology of the Black Coast, Antarctic Peninsula. In Craddock, C. ed. Antarctic Geoscience. Madison Univ. Wisconsin Press.

Rowley, P.D., Williams, P.L. and Schmidt, D.L. 1977. Geology of an Upper Cretaceous copper deposit in the Andean Province, Lassiter Coast, Antarctic Peninsula. Prof. Pap. U.S. geol. Surv., No. 984, 36pp.

Rowley P.D., Williams, P.L., Schmidt, D.L., Reynolds, R.L., Ford, A.B., Clark, A.H., Farrar, E. and McBride, S.L. 1975. Copper mineralization along the Lassiter Coast of the Antarctic Peninsula. Econ.Geol., 70, 982-992.

Saunders, A.D. 1982. Petrology and geochemistry of alkali basalts from Jason Peninsula, Oscar II coast. Br. Antarct. Surv. Bull., 55, 1-9.

Saunders, A.D. and Tarney, J. 1979. The geochemistry of basalt from a back-arc spreading centre in the East Scotia Sea. Geochem. Cosmochim. Acta. 43, 555-572.

Saunders, A.D. and Tarney, J, 1982. Igneous activity in the southern Andes and northern Antarctic Peninsula: a review. J. geol. Soc. Lond., 139, 691-700.

Saunders, A.D.,Tarney, J., Stern, C. and Dalziel, I.W.D. 1979. Geochemistry of Mesozoic marginal basin floor igneous rocks from southern Chile. Geol. Soc. Amer. Bull. 90, 237-258.

Saunders, A.D.,Tarney, J. and Weaver, S.D. 1980. Transverse geochemical variations across the Antarctic Peninsula: implications for the genesis of calc-alkaline magmas. Earth planet. Sci. Lett. 46 344-360.

Saunders, A.D., Tarney, J., Weaver, S.D. and Barker, P.F. 1982a. Scotia Sea floor: geochemistry of basalts from the Drake Passage and South Sandwich spreading centres. In Craddock C, ed. Antarctic Geoscience, Madison Univ. Wisconsin Press, 213-222.

Saunders, A.D.,Weaver, S.D. and Tarney, J. 1982b. The pattern of Antarctic Peninsula plutonism. In Craddock, C. ed Antarctic Geoscience Madison, Univ. Wisconsin Press, 263-274.

Scharnberger, C.K. and Scharon, L. 1972. Palaeomagnetism and plate tectonics of Antarctica. In Adie, R.J. ed. Antarctic Geology and Geophysics. Oslo, Universitetsforlaget, 843-947.

Sclater, J.G.,Bowin, C., Hey, R., Hoskins, H., Pierce, J., Phillips, J. and Tapscott, C. 1976. The Bouvet triple junction. J. geophys. Res., 81,1857-1869.

Sclater, J.G., Dick, H.,Norton, I.O. and Woodroffe, D. 1978. Tectonic structure and petrology of the Antarctic plate boundary near the Bouvet triple junction, Earth planet. Sci. Lett. 37, 393-400.

Scott, K.M. 1966. Sedimentology and dispersal pattern of a Cretaceous flysch sequence, Patagonian Andes, southern Chile. Bull. Am. Ass. Petrol. Geol. 50, 52-107.

Segoufin, J. 1978. Anomalies magnetiques mesozoignes dans le bassin de Mozambique. C.R. Acad. Sc. Paris, Ser. D., 287, 109-112.

Simpson, E.S.W.,Sclater, J.G., Parsons, B., Norton, I.O. and Meinke, K. 1979. Mesozoic magnetic lineations in the Mozambique Basin. Earth planet. Sci. Lett. 43, 260-4.

Simpson, P. and Griffiths, D.H. 1982. The structure of the South Georgia continental block. In Craddock, C. ed. Antarctic Geoscience, Madison Univ. Wisconsin Press, 185-192.

Skarmeta, J.M. 1974. Geologia de la region continental de Aisin entre los 45°-46° latitude sur. Santiago, Chile, Unpublished Ph.D. thesis, Universidad de Chile.

Smellie, J.L. 1981. A complete arc-trench system recognised in Gondwana sequences of the Antarctic Peninsula region. Geol. Mag. 118 139-159.

Smellie, J.L. and Clarkson, P.D. 1975. Evidence for pre-Jurassic subduction in western

Antarctica. Nature, 258, 701-02.
Smith, A.G. and Hallam, A, 1970. The fit of the southern Continents. Nature, 225, 139-144.

Stern, C.R. and Stroup, J.B. 1982. The petro-chemistry of the Patagonian batholith, Ultima Esperanza, Chile. In Craddock, C. ed. Antarctic Geoscience, Madison, Univ. Wisconsin. Press. 135-142.

Stern, C.R.,DeWit, M.J. and Lawrence, J.R. 1976. Igneous and metamorphic processes associated with the formation of Chilean ophiolites and their implication for ocean floor metamorphism, seismic layering and magmatism. J. geophys. Res.,81, 4370-4380.

Stern, C.R., Skewes, MA. and Duran, M. 1976. Calc-alkaline volcanism in southernmost Chile: Proc. of 1st Geologic Congress of Chile, 195-212.

Storey, B.C. and Mair, B.F. 1982. The composite floor of the Cretaceous back-arc basin of South Georgia. J. geol. Soc. Lond.139, 729-737.

Storey, B.C.,Mair, B.F. and Bell, C.M. 1977. The occurrence of Mesozoic ocean floor and ancient continental crust on South Georgia. Geol. Mag., 114, 203-208.

Stubbs, G.M. 1968. Geology of parts of the Foyn and Bowman Coasts, Graham Land. Unpublished Thesis, University of Birmingham, 245pp.

Suarez, M. 1976. Plate-tectonic model for south-ern Antarctic Peninsula and its relation to southern Andes. Geology, 4,211-14.

Suarez, M. and Pettigrew, T.H. 1976. An Upper Mesozoic island arc-back system in the southern Andes and South Georgia. Geol. Mag. 113, 305-328.

Tanner, P.W.G. 1982a. Geology of Shag Rocks, part of a continental block on the north Scotia Ridge, and possible regional correlations. Br. Antarct. Surv. Bull., 51, 125-136.

Tanner, P.W.G. 1982b. Geological evolution of South Georgia. In Craddock, C. ed. Antarctic Geoscience, Madison, Univ. Wisconsin Press, 167-176.

Tanner, P.W.G. and Macdonald, D.I.M., 1982. Models for the deposition and simple shear deformation of a turbidite sequence in the South Georgia portion of the southern Andes back-arc basin. J. geol. Soc. Lond. 139, 739-754.

Tanner, P.W.G., Pankhurst, R.J. and Hyden G.1982 Radiometric evidence for the age of the subduction complex in the South Orkney and South Shetland Islands, West Antarctica. J. geol. Soc. Lond. 139, 683-690.

Tarney, J.,Dalziel, I.W.D. and DeWit, M.J. 1976. Marginal basin 'Rocas Verdes' complex from S. Chile: A model for Archaean green-stone belt formation. In Windley, B.F., ed. Early History of the Earth: Wiley, London. 131-146.

Tarney, J., Saunders, A.D. and Weaver, S.D. 1977. Geochemistry of volcanic rocks from the island arcs and marginal basins of the Scotia Arc region. In Island arcs, deep

sea trenches and back-arc basins. Maurice Ewing Series 1. (ed. M. Talwani and W.C. Pitman III). American Geophysical Union. Washington, D.C. 367-378.

Tarney, J. Weaver, S.D., Saunders, A.D. Pankhurst, R.J. and Barker, P.F. 1982. Volcanic evolution of the Antarctic Peninsula and Scotia Arc. In. Thorpe R.S. ed. Orogenic Andesites. Wiley, London. 371-400.

Thorpe, R.S., Potts, P.J. and Francis, P.W. 1976. Rare earth data and petrogenesis of andesite from the North Chilean Andes, Contrib. Mineral. Petrol. 54. 65-78 .

Thomson, J.W. 1973. The geology of Powell, Christoffersen and Michelsen Inslands, South Orkney Islands. Br. Antarct. Surv. Bull. 33 and 34, 137-167.

Thomson, J.W. 1974. The geology of the South Orkney Islands: III, Coronation Island. Br Antarct. Surv. Sci. Rept. 86. 39pp.

Thomson, M.R.A. 1975. New palaeontological and lithological observations in the Legoupil Formation, north-west Antarctic Peninsula. Br. Antarct. Surv. Bull. 41 and 42. 169-185.

Thomson,M.R.A. 1982. Mesozoic palaeogeography of western Antarctica. In Craddock, C. ed. Antarctic Geoscience, Madison, Univ. Wisconsin Press. 331-337.

Thomson,M.R.A.,Tanner, P.W.G. and Rex, D.C. 1982. Fossil and Radiometric evidence for ages of deposition and metamorphism of sediment sequences on South Georgia. In Craddock, C. ed. Antarctic Geoscience, Madison, Univ, Wisconsin Press, 177-184.

Toksoz, M.N. and Bird, P. 1977. Formation and evolution of marginal basins and contin-ental plateaus. In Island Arcs, Deep Sea Trenches and Back-Arc Basins (eds. M. Talwani, and W.C. Pitman).. Maurice Ewing Series 1 379-393. Washington. American Geophysical Union.

Tomoda, Y., Kobayashi, K., Segawa, J.,Nomura, M., Kimora, K. and Saki, T. 1975. Linear magnetic anomalies in the Shikoku Basin, northeastern Philippine Sea. Jnl. Geomag. Geoelec. 28, 47-56.

Vlaar, N.J. and Wortel, M.J.R. 1976. Lithospheric ageing, instability and subduction. Tectonophysics 32, 331-351.

Watters, D.G. 1972. Geophysical Investigations of a section of the south Scotia Ridge. In. Adie, R.J. ed. Antarctic Geology and Geophysics. Universitetsforlaget, Oslo, 33-38.

Watts, A.B. and Weissel, J.K. 1975. Tectonic history of the Shikoku marginal basin. Earth planet. Sci. Lett. 25, 239-250

Watts, D.R. 1982. Potassium Argon and palaeo-magnetic results from King George Island, South Shetland Islands. In Craddock, C. ed. Antarctic Geoscience, Madison, Univ. Wisconsin Press, 255-261.

Weaver, S.D.,Saunders, A.D.,Pankhurst, R.J. and Tarney, J. 1979. A geochemical study of magmatism associated with the initial stages of back-arc spreading: the Quaternary volcanics of Bransfield Strait, South Shetland Islands. Contrib. Mineral. Petrol. 68, 151-169.

Weaver, S.D., Saunders, A.D. and Tarney, J. 1982. Mesozoic-Cainozoic volcanism in the South Shetland Islands and the Antarctic Peninsula: geochemical nature and plate tectonic significance. In Craddock, C. ed. Antarctic Geoscience, Madison, Univ. Wisconsin Press, 263-274.

Weissel, J.K., Hayes, D.E. and Herron, E.M. 1977. Plate tectonic synthesis: the displacements between Australia, New Zealand and Antarctica since the Late Cretaceous. Mar. Geol. 25, 231-277.

West, S.M. 1974. The geology of the Danco Coast, Graham Land, Br. Antarct. Surv. Sci. Rep. 84 58pp.

Williams, P.L., Schmidt, D.L., Plummer, C.C. and Brown, L.E. 1982. Geology of the Lassiter Coast area, Antarctic Peninsula: Preliminary report, in Adie, R.J. ed., Antarctic geology

and geophysics: Oslo, Universitetsforlaget, 143-148.

Windley, B.F. 1976. (ed) Early History of the Earth London, Wiley.

Winn, R.D. 1977. Submarine-fan turbidites and resedimented conglomerates in a Mesozoic arc-rear marginal basin in southern South America, in Stanley, D.J. and Kelling, G.K. eds. Sedimentation in submarine canyons, fans and trenches: Stroudberg, Pa., Dowden, Hutchinson and Ross.

Winn, R.D. Jr., 1978. Upper Mesozoic flysch of Tierra del Fuego and South Georgia Island: A sedimentological approach to lithosphere plate restoration. Geol. Soc. Amer. Bull. 89 533-547.

Winslow, M.A. 1982. The structural evolution of the Magallanes basin and neotectonics in the southernmost Andes. In Craddock, C. ed. Antarctic Geoscience, Madison, Univ. Wisconsin Press, 143-154.